W9-CEN-697

MICROCOSMOS

Four Billion Years of Evolution from Our Microbial Ancestors

Lynn Margulis and Dorion Sagan

Foreword by Dr. Lewis Thomas

A TOUCHSTONE BOOK
Published by Simon & Schuster
New York • London • Toronto • Sydney • Tokyo • Singapore

TOUCHSTONE
Simon & Schuster Building
Rockefeller Center
1230 Avenue of the Americas
New York, New York, 10020

First Touchstone Edition 1991
TOUCHSTONE and colophon are registered trademarks
of Simon & Schuster Inc.
Designed by Eve Kirch
Manufactured in the United States of America

1 3 5 7 9 10 8 6 4 2

1 3 5 7 9 10 8 6 4 2 Pbk.

Library of Congress Cataloging in Publication data
Margulis, Lynn, date.
Microcosmos: four billion years of evolution from our microbial ancestors/Lynn Margulis and Dorion Sagan; Foreword by Lewis Thomas.—1st Touchstone ed.
p. cm.
"A Touchstone book."
Reprint. Originally published: New York: Summit Books, © 1986.
Includes bibliographical references and index.
1. Evolution. 2. Microorganisms—Evolution. I. Sagan, Dorion, date. II. Title.
QH371.M28 1991
575—dc20 91–591
CIP

ISBN: 0-671-44169-8
ISBN: 0-671-74798-3 Pbk.

ACKNOWLEDGMENTS

In 1980 the colorful literary agent John Brockman, dressing like a fashion-conscious Italian gangster, came from New York to Boston to solicit a book from Lynn Margulis. Without that initial visit, and his continual encouragement, this book would never have been written. We are equally indebted to his partner, Katinka Matson, who has been nothing but helpful for the half decade this book was in gestation. We would like to thank Laszlo Meszoly who, at short notice, stippled the evocative scenes of life's evolution through time. We express our thanks to biology-watcher Lewis Thomas for writing the foreword and being an inspiration, world-traveler David Abram for sharing his insights into nature, and the late Theodore Sturgeon, whose science fiction story "Microcosmic God" is paraphrased within these pages. We also are deeply grateful to our friends, family, editors, publishers, and colleagues Morris Alexander, David Bermudes, Robert Boynton, Jack Corliss, Geoff Cowley, Eileen Crist, W. Ford Doolittle, Ann Druyan, Betsey Dexter Dyer, Stephen

Jay Gould, Bruce Gregory, Ricardo Guerrero, James Hallgring, Stephanie Hiebert, Donald Johanson, Geraldine Kline, Edmond LeBlanc, James Lovelock, Heinz Lowenstam, David Lyons, Lorraine Olendzenski, Jennifer Margulis, Zachary Margulis, Kelly McKinney, Philip and Phylis Morrison, Elaine Pagels, John Platt, Carl Sagan, Jeremy Sagan, Marjorie MacLean, Arthur Samuelson, Nathan Shafner, James Silberman, William Solomon, John Stolz, William Irwin Thompson, Paul Trachtman, and Peggy Tsukahira.

We are sorry that our colleague and friend Elso S. Barghoorn, who helped us with this book in its early stages, did not live to see it completed.

Most of the scientific research on which some of this text is based has been supported by the planetary biology office of NASA and the Boston University graduate school. We are also grateful to The Lounsbery Foundation for continuing support of this work. Many of the conclusions we have drawn are based on research and discussion available in the scientific literature. Of course we owe a major debt to the many unmentioned authors and scientists whose work provides a basis for our narrative.

Lynn Margulis and Dorion Sagan
February 1986/1990
Boston/Amherst

To Morris Alexander,
father and grandfather,
and his love of life

CONTENTS

TABLES

FOREWORD

by
Lewis Thomas, M.D., President Emeritus,
Memorial Sloan-Kettering Cancer Center

IT is on occasion the function of a foreword to provide the reader with advance notice of what he or she may be in for. In the case of this book, unless the reader has been keeping in close touch with quite recent events in microbiology, paleontology and evolutionary biology, what he or she is in for is one great surprise after another, even possibly one shock after another. This is a book about the inextricable connectedness of all creatures on the planet, the beings now alive and all the numberless ones that came before. Margulis and Sagan propose here a new way of looking at the world, different from the view we mostly shared a few decades back. The new view is based on solid research, done for different reasons by many scientists in laboratories all around the earth. Brought together and linked, their findings lead to the conclusion that separateness is out of the question in Nature. The biosphere is all of a piece, an immense, integrated living system, an organism.

I remember attending a series of seminars on a university

campus long ago, formally entitled "Man's Place in Nature." Mostly, it had to do with how man can fix Nature up, improving it so that the world's affairs might move along more agreeably: how to extract more of the Earth's energy resources, how to preserve certain areas of wilderness for our pleasure, how to avoid polluting the waterways, how to control the human population, things of that order. The general sense was that Nature is a piece of property, an inheritance, owned and operated by mankind, a sort of combination park, zoo and kitchen garden.

This is still the easy way to look at the world, if you can keep your mind from wandering. Surely, we have had the appearance of a dominant species, running the place, for almost the entire period of our occupancy. At the beginning, perhaps, we were fragile, fallible creatures, just down from the trees with nothing to boast of beyond our apposable thumbs and our exaggerated frontal lobes, hiding in caves and studying fire. But we took over, and now we seem to be everywhere, running everything, pole to pole, mountain peaks to deep sea trenches, colonizing the moon and eyeing the solar system. The very brains of the Earth. The pinnacle of evolution, the most stunning of biological successes, here to stay forever.

But there is another way to look at us, and this book is the guide for that look. In evolutionary terms, we have only just arrived. There may be younger species than ours, here and there, but none on our scale, surely none so early on in their development. We cannot trace ourselves back more than a few thousand years before losing sight of what we think of as the real human article, language-speaking, song-singing, tool-making, fire-warming, comfortable, war-making mankind. As a species, we are juvenile, perhaps just beginning to develop, still learning to be human, an immature child of a species. And vulnerable, error-prone still, at risk of leaving only a thin layer of radioactive fossils.

One thing we need to straighten out in aid of our perspective is our lineage. We used to believe that we arrived *de novo,* set in place by the Management, maybe not yet dressed but ready anyway to name all the animals. Then, after Darwin, we had to face up to the embarrassment of having apes somewhere in the family tree, with chimps as cousins.

Many children go through a painful period in early adolescence when they are uncomfortable about their parents, wishing them to be different, more like the parents of families down the street. There is nothing really shameful about having odd-looking hominids as parents, but still most of us would prefer, given the choice, to track our species back to pure lines of kings and queens, stopping there and looking no further.

But now look at our dilemma. The first of us, the very first of our line, appeared sometime around 3.5 billion years ago, a single bacterial cell, the Ur-ancestor of all the life to come. We go back to *it,* of all things.

Moreover, for all our elegance and eloquence as a species, for all our massive frontal lobes, for all our music, we have not progressed all that far from our microbial forebears. They are still with us, part of us. Or, put it another way, we are part of them.

Once faced up to, it is a grand story, a marvelous epic, still nowhere near its end. It is nothing less than the saga of the life of the planet.

Lynn Margulis has been spending most of her professional life studying the details of the story and has added significant details from her own scientific research. Now, she and Dorion Sagan have put it all together, literally, in this extraordinary book, which is unlike any treatment of evolution for a general readership that I have encountered before. It is a fascinating account of what is by far the longest stage in the evolution of the biosphere, the 2.5 billion year stretch of time in which our microbial ancestors, all by themselves, laid out most of the rules and regulations for interliving, habits we humans should be studying now for clues to our own survival.

Most popular accounts of evolution and its problems start out just a few hundred million years ago, paying brief respects to the earliest forms of multicellular organisms and then moving quickly to the triumphant invention of vertebrate forms, making it seem as though all the time that went before was occupied by "primitive" and "simple" cells doing nothing but waiting around for the real show to begin. Margulis and Sagan fix this misapprehension of the real facts of life, demonstrating that the earliest bacteria learned almost everything there is to know about living in a system, and they are, principally, what we know today.

Perhaps we have had a shared hunch about our real origin longer than we think. It is there like a linguistic fossil, buried in the ancient root from which we take our species' name. The word for earth, at the beginning of the Indoeuropean language thousands of years ago (no one knows for sure how long ago) was *dhghem*. From this word, meaning simply *earth* came our word *humus*, the handiwork of soil bacteria. Also, to teach us the lesson, *humble*, *human*, and *humane*. There is the outline of a philological parable here; some of the details are filled in by this book.

INTRODUCTION: THE MICROCOSM

WHEN people look at life on Earth, it is easy to think we are supreme. The power of consciousness, of our society and our technical inventions, has made us think we are the most advanced form of life on the planet. Even the great blackness of space seen does not humble us. We view space as a no man's land to penetrate and conquer as we believe we have conquered the Earth.

Life on Earth has traditionally been studied as a prologue to humans: "lower" forms of life lacking intelligence preceded us and we now stand at the pinnacle of evolution. Indeed, so godlike do we consider ourselves that we may think we are taking evolution into our own hands by manipulating DNA, the mainspring of life, according to our own design. We study the microcosm—the age-old world of microorganisms—to discover life's secret mechanisms so that we can take better control, perhaps even "perfect" ourselves and the other living things on the Earth.

But during the past three decades, a revolution has taken

place in the life sciences. Fossil evidence of primeval microbial life, the decoding of DNA, and discoveries about the composition of our own cells have exploded established ideas about the origins of life and the dynamics of evolution on Earth.

First, they have shown the folly of considering people as special, apart and supreme. The microscope has gradually exposed the vastness of the microcosm and is now giving us a startling view of our true place in nature. It now appears that microbes—also called microorganisms, germs, bugs, protozoans, and bacteria, depending on the context—are not only the building blocks of life, but occupy and are indispensable to every known living structure on the Earth today. From the paramecium to the human race, all life forms are meticulously organized, sophisticated aggregates of evolving microbial life. Far from leaving microorganisms behind on an evolutionary "ladder," we are both surrounded by them and composed of them. Having survived in an unbroken line from the beginnings of life, all organisms today are equally evolved.

This realization sharply shows up the conceit and presumption of attempting to measure evolution by a linear progression from the simple—so-called lower—to the more complex (with humans as the absolute "highest" forms at the top of the hierarchy). As we shall see, the simplest and most ancient organisms are not only the forebears and the present substrate of the Earth's biota, but they are ready to expand and alter themselves and the rest of life, should we "higher" organisms, be so foolish as to annihilate ourselves.

Next, the view of evolution as chronic bloody competition among individuals and species, a popular distortion of Darwin's notion of "survival of the fittest," dissolves before a new view of continual cooperation, strong interaction, and mutual dependence among life forms. Life did not take over the globe by combat, but by networking. Life forms multiplied and complexified by co-opting others, not just by killing them.

Because we cannot see the microcosm with the unaided eye, we tend to discount its significance. Yet of the three-and-a-half billion years that life has existed on Earth, the entire history of human beings from the cave to the condominium represents far less than one percent. Not only did life originate on earth very early in its history as a planet, but for the first full two billion years, Earth was inhabited solely by bacteria.

In fact, so significant are bacteria and their evolution that the fundamental division in forms of life on Earth is not that between plants and animals, as is commonly assumed, but between prokaryotes—organisms composed of cells with no nucleus, that is, bacteria—and eukaryotes—all the other life forms.[1] In their first two billion years on Earth, prokaryotes continuously transformed the Earth's surface and atmosphere. They invented all of life's essential, miniaturized chemical systems—achievements that so far humanity has not approached. This ancient high *bio*technology led to the development of fermentation, photosynthesis, oxygen breathing, and the removal of nitrogen gas from the air. It also led to worldwide crises of starvation, pollution, and extinction long before the dawn of larger forms of life.

These staggering events early in life's history came about by the interaction of at least three recently discovered dynamics of evolution. The first is the remarkable orchestrating abilities of DNA. Identified as the heredity-transmitting substance in 1944 by Oswald T. Avery, Colin MacLeod, and

Maclyn McCarty, DNA's code was cracked in the 1960s after its method of replication was revealed by James Watson and Francis Crick in 1953. Governed by DNA, the living cell can make a copy of itself, defying death and maintaining its identity by reproducing. Yet by also being susceptible to mutation, which randomly tinkers with identity, the cell has the potential to survive change.

A second evolutionary dynamic is a sort of natural genetic engineering. Evidence for it has long been accumulating in the field of bacteriology. Over the past fifty years or so, scientists have observed that prokaryotes routinely and rapidly transfer different bits of genetic material to other individuals. Each bacterium at any given time has the use of accessory genes, visiting from sometimes very different strains, which perform functions that its own DNA may not cover. Some of the genetic bits are recombined with the cell's native genes; others are passed on again. Some visiting genetic bits can readily move into the genetic apparatus of eukaryotic cells (such as our own) as well.

These exchanges are a standard part of the prokaryotic repertoire. Yet even today, many bacteriologists do not grasp their full significance: that as a result of this ability, all the world's bacteria essentially have access to a single gene pool and hence to the adaptive mechanisms of the entire bacterial kingdom. The speed of recombination over that of mutation is superior: it could take eukaryotic organisms a million years to adjust to a change on a worldwide scale that bacteria can accommodate in a few years. By constantly and rapidly adapting to environmental conditions, the organisms of the microcosm support the entire biota, their global exchange network ultimately affecting every living plant and animal. Human beings are just learning these techniques in the science of genetic engineering, whereby biochemicals are produced by introducing foreign genes into reproducing cells. But prokaryotes have been using these "new" techniques for billions of years. The result is a planet made fertile and inhabitable for larger forms of life by a communicating and cooperating worldwide superorganism of bacteria.

Far-reaching as they are, mutation and bacterial genetic transfer alone do not account for the evolution of all the life forms on the earth today. In one of the most exciting discoveries of modern microbiology, clues to a third avenue of change appeared in the observation of mitochondria—tiny membrane-wrapped inclusions in the cells of animals, plants, fungi, and protists alike. Although they lie outside the nucleus in modern cells, mitochondria have their own genes composed of DNA. Unlike the cells in which they reside, mitochondria reproduce by simple division. Mitochondria reproduce at different times from the rest of the cell. Without mitochondria, the nucleated cell, and hence the plant or animal, cannot utilize oxygen and thus cannot live.

Subsequent speculation brought biologists to a striking scenario: The descendants of the bacteria that swam in primeval seas breathing oxygen three billion years ago exist now in our bodies as mitochondria. At one time, the ancient bacteria had combined with other microorganisms. They took up residence inside, providing waste disposal and oxygen-derived energy in return for food and shelter. The merged organisms went on to evolve into more complex oxygen-breathing forms of life. Here, then, was an evolutionary mechanism more sudden than mutation: a symbiotic alliance that becomes permanent. By creating organisms that are not simply the sum of their symbiotic parts—but something more like the sum of all the possible combinations of their parts—such alliances push developing beings into uncharted realms. Symbiosis, the merging of organisms into new collectives, proves to be a major power of change on Earth.[2]

As we examine ourselves as products of symbiosis over billions of years, the supporting evidence for our multimicrobe ancestry becomes overwhelming. Our bodies contain a veritable history of life on Earth. Our cells maintain an environment that is carbon- and hydrogen-rich, like that of the Earth when life began. They live in a medium of water and salts like the composition of the early seas. We became who we are by the coming together of bacterial partners in a watery environment. Although the evolutionary dynamics of DNA, genetic transfer, and symbiosis were not discovered until almost a century after Charles Darwin's death in 1882, he had the shrewdness to write, "We cannot fathom the marvellous complexity of an organic being; but on the hypothesis here advanced this complexity is much increased. Each living creature must be looked at as a microcosm—a little universe, formed of a host of self-propagating organisms, inconceivably minute and as numerous as the stars in heaven."[3] The strange nature of this little universe is what this book is about.

The detailed structure of our cells betrays the secrets of their ancestors. Electron microscopic images of nerve cells from all animals reveal numerous conspicuous "microtubules." The waving cilia in the lining of our throats and the whipping tail of the human sperm cell both have the same unusual "telephone dial" arrangement of microtubules as do the cilia of ciliates, a group of successful microbes including more than eight thousand different species. These same microtubules appear in all cells of plants, animals, and fungi each time the cells divide. Enigmatically, the microtubules of dividing cells are made of a protein nearly identical to one found in our brains; and this is a protein exceedingly similar to some of those found in certain fast-moving bacteria shaped like corkscrews.

These and other living relics of once-separate individuals, detected in a variety of species, make it increasingly certain that all visible organisms evolved through symbiosis, the coming together that leads to physical interdependence and the permanent sharing of cells and bodies. Although, as we shall see, some details of the bacterial origin of mitochondria, microtubules, and other cell parts are hard to explain, the general outline of how evolution can work by symbiosis is agreed upon by those scientists who are familiar with the lifestyles of the microcosm.

The symbiotic process goes on unceasingly. We organisms of the macrocosm continue to interact with and depend upon the microcosm, as well as upon each other. Certain families of plants (such as the pea family, including peas, beans, and their relatives such as clover and vetch) cannot live in nitrogen-poor soil without the nitrogen-fixing bacteria in their root nodules, and we cannot live without the nitrogen that comes from such plants. Neither cows nor termites can digest the cellulose of grass and wood without communities of microbes in their guts. Fully ten percent of our own dry body weight consists of bacteria, some of which, although they are not a congenital part of our bodies, we can't live without. No mere quirk of nature, such coexistence is the stuff of evolution itself. Let evolution continue a few million years more, for example, and those microorganisms producing vitamin B_{12} in our intestines may become parts of our own cells. An aggregate of specialized cells may become an organ. The union of once-lethal bacteria with amoebae, creating over time a new species of hybrid amoeba, has even been witnessed in the laboratory.

This revolution in the study of the microcosm brings before us a breathtaking view. It is not preposterous to postulate that the very consciousness that enables us to probe the workings of our cells may have been born of the concerted capacities of millions of microbes that evolved symbiotically to become the human brain. Now, this consciousness has led us to tinker with DNA and we have begun to tap in to the ancient process of bacterial genetic transfer. Our ability to make new kinds of life can be seen as the newest way in which organic memory—life's recall and activation of the past in the present—becomes more acute. In one of life's giant, self-referential loops, changing DNA has led to the consciousness that enables us to change DNA. Our curiosity, our thirst to know, our enthusiasm to enter space and spread ourselves and our probes to other planets and beyond represents part of the cutting edge of life's strategies for expansion that began in the microcosm some three-and-a-half billion years ago. We are but reflections of an ancient trend.

From the first primordial bacteria to the present, myriads of symbiotically formed organisms have lived and died. But the microbial common denominator remains essentially unchanged. Our DNA is derived in an unbroken sequence from the same molecules in the earliest cells that formed at the edges of the first warm, shallow oceans. Our bodies, like those of all life, preserve the environment of an earlier Earth. We coexist with present-day microbes and harbor remnants of others, symbiotically subsumed within our cells. In this way, the microcosm lives on in us and we in it.

Some people may find this notion disturbing, unsettling. Besides popping the overblown balloon that is our presumption of human sovereignty over the rest of nature, it challenges our ideas of individuality, of uniqueness and independence. It even violates our view of ourselves as discrete physical beings separated from the rest of nature. To think of ourselves and our environment as an evolutionary mosaic of microscopic life evokes imagery of being taken over, dissolved, annihilated. Still more disturbing is the philosophical conclusion we will reach later, that the possible cybernetic control of the Earth's surface by unintelligent organisms calls into question the alleged uniqueness of human intelligent consciousness.

Paradoxically, as we magnify the microcosm to find our origins, we appreciate sharply both the triumph and the insignificance of the individual. The smallest unit of life—a single bacterial cell—is a monument of pattern and process unrivaled in the universe as we know it. Each individual that grows, doubles its size, and reproduces is a great success story. Yet just as the individual's success is subsumed in that of its species, so is the species subsumed in the global network of all life—a success of an even greater order of magnitude.

It is tempting, even for scientists, to get carried away by success stories. From the disciples of Darwin to today's genetic engineers, science has popularized the view that humans are at the top rung of Earth's evolutionary "ladder" and that with technology we have stepped outside the framework of evolution. Some eminent and sophisticated scientists, such as Francis Crick in his book, *Life Itself*, write that life in general and human consciousness in particular are so miraculous that they couldn't be earthly at all, but must have originated elsewhere in the universe.[4] Others still believe that humans are a product of a fatherly "higher intelligence"—the children of a divine patriarch.

This book was written to show that these views underestimate the Earth and the ways of nature. There is no evidence that human beings are the supreme stewards of life on Earth, nor the lesser offspring of a superintelligent extraterrestrial source. But there is evidence to show that we are recombined from powerful bacterial communities with a multibillion-year-old history. We are a part of an intricate network that comes from the original bacterial takeover of the Earth. Our powers of intelligence and technology do not belong specifically to us but to all life. Since useful attributes are rarely discarded in evolution it is likely that our powers, derived from the microcosm, will endure in the microcosm. Intelligence and technology, incubated by humankind, are really the property of the microcosm. They may well survive our species in forms of the future that lie beyond our limited imaginations.

AUTHORS' PREFACE

What is the relationship between humans and Nature? The Linnaean, or scientific, name of our own species is *Homo sapiens sapiens*—"Man, the wise, the wise." But, as a humble proposal or wisecrack, we suggest that humanity be rechristened *Homo insapiens*—"Man, the unwise, the tasteless." We love to think we are Nature's rulers—"Man is the measure of all things," said Protagoras 2,400 years ago—but we are less regal than we imagine. *Microcosmos: Four Billion Years of Evolution from Our Microbial Ancestors* (New York: Summit Books, 1986) strips away the gilded clothing that serves as humanity's self-image to reveal that our self-aggrandizing view of ourselves is no more than that of a planetary fool.

Humans have long been the planetary or biospheric equivalent of Freud's ego, which "plays the ridiculous role of the clown in the circus whose gestures are intended to persuade the audience that all the changes on the stage are brought about by his orders." We resemble such a clown except that, unlike him, our egotism concerning our own importance for

Nature is often humorless. Freud continues, "But only the youngest members of the audience are taken in by him."[5] Perhaps human gullibility regarding planetary ecology is also a function of our youth—our collective immaturity as one of many species sharing the Earth. But even if we are Nature's brilliant child, we are not that scientific conceit, "the most highly evolved species." The human "emperor," from the revisionary perspective of *Microcosmos,* and in the humble opinion of its authors, is wearing no clothes.

A recent forum in *Harper's Magazine,* titled "Only Man's Presence Can Save Nature,"[6] exemplifies humanity's typically grandiose, almost solipsistic, view of itself. Atmospheric chemist James Lovelock speaks of the relationship between humans and Nature as an impending "war"; ecofundamentalist Dave Foreman declares that, far from being the central nervous system or brain of Gaia, we are a cancer eating away at her; while University of Texas Professor of Arts and Humanities Frederick Turner transcendentally assures us that humanity is the living incarnation of Nature's billion-year-old desire. We would like to take all these views to task. In medieval times an interesting prop of the jesting Fool, besides glittering jeweled bauble and wooden knife, was the globe. Picture this figure—capped and belled Fool, ear flaps a-dangling as he handles a mock Earth—for a more festive, if no less true, summary of how things stand between *Homo sapiens* and Nature.

Through Plato, Socrates speaks of the folly of inscribing one's opinions: although your views may change, your words as committed to paper remain. Socrates at least did not write, and what he knew, first and foremost, was that he did not know. We, however, did write. Reversing the usual inflated view of humanity, we wrote of *Homo sapiens* as a kind of latter-day permutation in the ancient and ongoing evolution of the smallest, most ancient, and most chemically versatile inhabitants of the Earth, namely bacteria. We wrote that the physiological system of life on Earth, Gaia, could easily survive the loss of humanity, whereas humanity would not survive apart from that life. *Microcosmos* received generally favorable reviews, but was criticized on several scores, most vehemently for our cavalier attitude toward our own species. We outraged some with the implication that even nuclear war would not be a total apocalypse, since the hardy bacteria underlying life on a planetary scale would doubtless survive it. Unlike spoken words floating off noncommittally into the fickle winds of opinion, our words as hard symbols on paper sat, as here they sit—obstinately confronting us with dogma and didacticism instead of what otherwise might have been merely a provisional opinion. Happily, though, the occasion of the paperback reprinting of *Microcosmos* offers us an opportunity, if not to rewrite and revise, at least to reflect on the book and its main concerns.

Much has occurred, in science and in the world, in the half decade since the hardcover first appeared. In "The Symbiotic Brain" (Chapter 9) we detailed the speculation that the sperm tails of men, which propel sperm to the eggs of women, evolved through symbiosis. We claimed that sperm tails and oviduct undulipodia (among other subvisible structures) derived from spirochete bacteria that became ancestral cell "whips." In 1989 three Rockefeller University scientists published an arcane report of a new special cell DNA.[7] Although the research has been challenged, their discovery of "kinetosome" DNA, outside the cell nucleus and tightly packed at the base of each cell whip (undulipodium), is the single most important scientific advance for the symbiotic theory of cell evolution since the 1953 discovery of DNA itself. *Microcosmos,* in contrast to the usual view of neo-Darwinian evolution as an unmitigated conflict in which only the strong survive, more than ever encourages exploration of an essential alternative: a symbiotic, interactive view of the history of life on Earth. And although we would be foolish to propose that competitive power struggles for limited space and resources play no role in evolution, we show how it is equally foolish to overlook the crucial importance of physical association between organisms of different species, symbioses, as a major source of evolutionary novelty. And during the last half decade events and moods have tended to underscore the importance of symbiosis and cooperation far beyond the microworld of evolving bacteria.

As symbolized by the deconstruction of the Berlin Wall and the end of the Cold War, it is folly not to extend the lessons of evolution and ecology to the human and political realm. Life is not merely a murderous game in which cheating and killing insure the injection of the rogue's genes into the next generation, but it is also a symbiotic, cooperative venture in which partners triumph. Indeed, despite the belittling of humanity that naturally occurs when one looks at "*Homo sapiens sapiens*" from a planetary perspective of billions of years of cell evolution, we can rescue for ourselves some of our old evolutionary grandeur when we recognize our species not as lords but as partners: we are in mute, incontrovertible partnership with the photosynthetic organisms that feed us, the gas producers that provide oxygen, and the heterotrophic bacteria and fungi that remove and convert our waste. No political will or technological advance can dissolve that partnership.

Another sign of this distinct sort of deserved grandeur is our involvement in a project that may well outlast our species as we know it: the introduction of biospheres[8] to other planets and to outer space. These expanding activities resemble nothing so much as the reproduction of the planetary living system—the truly physiologically behaving nexus of all life on Earth. The expansion and reproduction of the biosphere, the production of materially closed, energetically open ecosystems on the Moon, Mars, and beyond, depends upon humanity in its widest sense as a planetary-technological phenomenon. David Abram, a philosopher at SUNY–Stony Brook, has spoken of humanity "incubating" technology. A selfish attitude and an exaggerated sense of our own importance may have spurred the augmentation of technology and human population at the expense of other organisms. Yet now, after the "incubation phase," the Gaian meaning of technology reveals itself: as a human-mediated but not a human phenomenon, whose applications stand to expand the influence of all life on Earth, not just humanity.

In *Microcosmos* we retrace evolutionary history from the novel perspective of the bacteria. Bacteria, single and multicellular, small in size and huge in environmental influence, were the sole inhabitants of Earth from the inception of life nearly four billion years ago until the evolution of cells with nuclei some two billion years later. The first bacteria were anaerobes: they were poisoned by the very oxygen some of them produced as waste. They breathed in an atmosphere that contained energetic compounds like hydrogen sulfide and methane. From the microcosmic perspective, plant life and animal life, including the evolution of humanity, are recent, passing phenomena within a far older and more fundamental microbial world. Feeding, moving, mutating, sexually recombining, photosynthesizing, reproducing, overgrowing, predacious, and energy-expending symbiotic microorganisms preceded all animals and all plants by at least two billion years.

What is humanity? The Earth? The relationship of the two, if in fact they are two? *Microcosmos* approaches these large questions from the particular perspective of a planet whose evolution has been largely a bacterial phenomenon. We believe this formerly slighted perspective is a highly useful, even essential, compensation required to balance the traditional anthropocentric view which flatters humanity in an unthinking, inappropriate way. Ultimately we may have overcompensated. In the philosophical practice known as deconstruction, powerful hierarchical oppositions are dismantled by a dual process Jacques Derrida caricatures or characterizes as "reversal and displacement." This process is at work in *Microcosmos:* humanity is deconstructed as the traditional hierarchy—recently evolved humans on top, evolutionarily older "lower" organisms below—is reversed. *Microcosmos* removes man from the summit, showing the immense ecological and evolutionary importance of the lowest of the small organisms, bacteria. But from the view of deconstructive practice, *Microcosmos,* which reverses the hierarchical opposition, does not take the next step of displacement: man is taken off the top of Nature only to be put on the bottom. What ultimately must be called into question is not the position assumed by humans in the opposition Man/Nature but the oppositional distortions imposed by the hierarchy itself. (A more parochial matter for deconstruction, apparently of interest to Derrida himself, is the hierarchy humanity/animality.) If we were to entirely rewrite *Microcosmos,* we might try to redress the naïveté of this inversion, which—like turning the king into a fool—upsets our conventions, but only in a preliminary fashion without truly dismantling them. Nearly all our predecessors assumed that humans have some immense importance, either material or transcendental. We picture humanity as one among other microbial phenomena, employing *Homo insapiens* as a nickname to remind ourselves to stave off the recurring fantasy that people master (or can master) Gaia. The microbial view is ultimately provisional; there is no absolute dichotomy between humans and bacteria. *Homo insapiens,* our more humbling name, seems more fitting, somehow more "Socratic." At least we know, it says, that we do not know.

The *Harper's* debate presented a diversity of characterizations of the relationship between "Man" and "Nature." And despite the title, "Only Man's Presence Can Save Nature," the editors dutifully informed us that one of the most significant contributions to the debate on humanity's status is that "Nature has ended." In *Microcosmos* we take a stance against the division of humans beings from the rest of "Nature." People are neither fundamentally in conflict with nor essential to the global ecosystem. Even if we accomplish the extraterrestrial expansion of life, it will not be to the credit of humanity *as humanity.* Rather it will be to the credit of humanity as a symbiotically evolving, globally interconnected, technologically enhanced, microbially based system. Given time, raccoons might also manufacture and launch their ecosystems as space biospheres, establishing their bandit faces on other planets as the avant-garde of Gaia's strange and seedlike brood. Maybe not black-and-white raccoons, but diaphanous nervous-system fragments of humanity, evolved beyond recognition as the organic components of reproducing machines, might survive beyond the inevitable explosion and death of the sun. Our microcosmic portrayal of *Homo sapiens sapiens* as a kind of glorified sludge has the merit of reminding us of our bacterial ancestry and our connections to a still largely bacterial biosphere.

An old metaphysical prejudice, a thinly disguised axiom of western philosophy, is that human beings are radically separate from all other organisms. Descartes held that nonhuman animals lacked souls. For centuries scientists have suggested that thought, language, tool use, cultural evolution, writing, technology—something, *anything*—unequivocally distinguishes people from "lower" life forms. As recently as 1990 nature writer William McKibben wrote, "In our modern minds nature and human society are separate things . . . this separate nature . . . is quite real. It is fine to argue, as certain poets and biologists have, that we must learn to fit in with nature, to recognize that we are but one species among many. . . . But none of us, on the inside, quite believe it."[9] Perhaps this anthropocentric self-glorification spurred our ancestors on, gave them the confidence to be "fruitful and multiply"—to rush to the very brink we are on now of a punctuated change in global climate, accompanied by mass extinctions and a shift in the Gaian "geophysiology."

It is usually thought that Darwin, by presenting evidence for the theory of evolution by natural selection, dramatically knocked the pedestal out from under the feet of humanity, undermining the case for God, leaving us uncomfortably in the company of other animals by broadcasting the taboo secret of our apish origins. The Darwinian revolution has often been compared to that of Copernicus, who showed that the Earth is not the center of the Universe but merely a dust speck in the corner of our galactic Milky Way cobweb. From a philosophical point of view, however, far from the Darwinian revolution destroying our special relationship as the unique life form, as a chosen species made in God's image and with connections to saints and angels, what seems to have happened in the wake of the Darwinian revolution is that we, *Homo sapiens sapiens*—man, the wise, the wise—have come to replace God. No longer are we junior partners, second in command. Darwinism may have destroyed the anthropomorphic deity of traditional religion, but instead of humbling us into awareness of the protoctists and all other sibling life forms (the plants, fungi, bacteria, and other animals), it rendered us greedy to assume God's former place. We put ourselves in the self-assumed position of divine rulers over life on Earth, ambitiously devising planet-scale technologies and, in short, engineering the world.

Somewhat surprisingly to those not versed in the ways of feedback, this self-serving attitude of human glorification at the expense of other species no longer serves us. Our extreme self-centeredness and hyperpopulation of the planet have brought on wholesale ecological carnage, the greatest threat of which is to ourselves. The traditional religious perspective—kept alive, as we have seen, even inside secular Darwinism—is that human beings are separate, unique, better. This is the attitude of ecological arrogance. The perspective of *Microcosmos* differs in that it is a deep-ecology, a particular variety of "green" perspective. Referring now to Lewis Thomas's tracing of the early word *human* in the Foreword, *Microcosmos* tries to develop an attitude of ecological *humility*. Retelling the story of life from the vantage point of microbes, *Microcosmos* diametrically inverts the usual hierarchy: indeed, by claiming that the planetary system of life has no essential need for man, that humanity is a temporary pointillist epiphenomenon of the essential and anciently recombining microorganisms, we may have overstated, exaggerated the case. The problem with the reversal that places microbes on top and people underneath is that dichotomization—important versus unimportant, essential versus unessential—remains. Woody Allen once said that he always put his wife *under* a pedestal. Confronting our ecological arrogance does not solve the problem of the pedestal: it is still assumed that one organism is better, higher, or "more evolved" than the other. To deconstruct our destructive attitude of ecological arrogance, it is necessary to put ourselves down. Once we recognize our energetic and chemical intercourse with other species, however, and the nonnegotiability of our connections with them, we must remove the pedestal altogether.

In tandem with its attempt to carry to the limits Darwin's "Copernican" revolution, *Microcosmos* stresses the symbiotic history of life. Since the publication of the hardcover, more striking evidence has accumulated to show that symbiosis, the living together and sometimes merging of different species of organisms, has been crucial to the evolution of life forms on Earth. The most important examples of symbiosis—the chloroplasts of all plants and the mitochondria of all animals, both of which were formerly independent bacteria—are of course well detailed in *Microcosmos*. But symbiosis now appears to be particularly good as an explanation of "jumps" in evolution that have momentous ecological importance. Submarine fishes, luminously spotlighting the blackest of waters, may have evolved into myriads of kinds, spurred on by eye-patch, esophageal, or anal light organs harboring glowing symbiotic bacteria.[10] Different symbioses of fish and beetles with glow-in-the-dark bacteria abound.

Another example of recent symbiosis research suggests that the green algal transition to land plants resulted from a merging of the genomes (genetic material) of a fungus with some aquatic green alga ancestor. Lichens are well-known products of symbioses. All lichens are fungi symbiotic with cyanobacteria or fungi symbiotic with green algae. The two types of life—photosynthesizer and consumer—intertwine to form a novel green low-lying plantlike organism with remarkable longevity—the lichen. The amazing capacity of lichens to proliferate on the bare face of rocks depends on the symbiosis, the equally combined fungal and photosynthetic partners that comprise the lichen entity. The newest twist is that vascular plants—including herbs and shrubs and all trees—may originally have been "inside-out lichens." Their evolution may have involved a new partnership between widely differing species from different kingdoms of life. If Professor Peter Atsatt's theory[11] is correct, then the interactive venture between two kinds of organisms, fungi and protoctistan green algae, accounted not for the appearance of some minor entity in the backwoods of evolution but for the momentous evolution of the Kingdom Plantae, the woods themselves.

The illusion of the independence of humans from Nature is dangerous ignorance. An unbroken continuum of life exists now as it has since life's inception—through Darwinian time (four billion years) and Vernadskian space (a twenty-five kilometer ring, extending ten kilometers down to the abyss and fifteen to the top of the troposphere). Inside this living system we are all embedded: to escape it is tantamount to death. Emily Dickinson, noting "what mystery pervades a well,"[12] charmingly described us and Nature. It is fitting to cite her prior to the descent into the microcosm:

> But nature is a stranger yet;
> The ones that cite her most
> Have never passed her haunted house,
> Nor simplified her ghost.
> To pity those that know her not
> Is helped by the regret
> That those who know her,
>   know her less
> The nearer her they get.[13]

—Dorion Sagan and Lynn Margulis
January 1991

Microcosmos

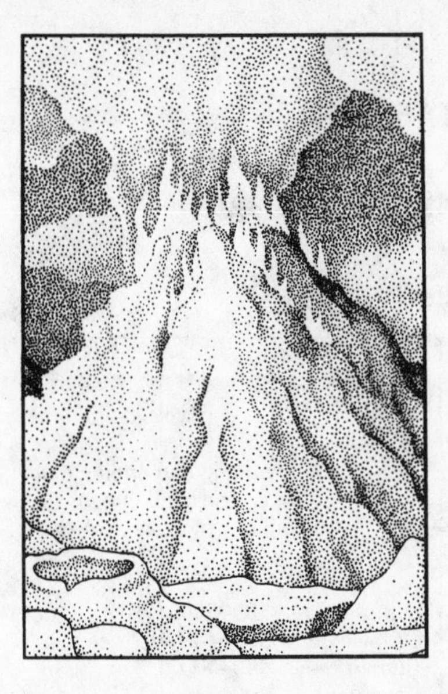

CHAPTER 1

Out of the Cosmos

FROM the moment we consider origins on a cosmic scale, the view of ourselves as a part—a miniscule part—of the universe is thrust upon us. For the very atoms that compose our bodies were created not, of course, when we were conceived, but shortly after the birth of the universe itself.

It is a known astrophysical fact that most stars in the sky are shooting away from each other at tremendous speeds. If we reverse this trend in our minds we come up with the so-called Big Bang, the hypothetical release of all the energy, matter, and antimatter in existence. Like any other look into what Shakespeare called "the dim backward and abysm of time," we must not mistake our best guesses or relatively straight-line extrapolations of present conditions into the past for the literal truth. Slight alterations in the most minor assumptions can lead to major distortions when magnified over the 15,000 million year time span that is the purported age of the present universe. Nonetheless, such extrapolations yield the best picture we have of the cosmos which preceded

TABLE 1
GEOLOGICAL TIME SCALE*
(in millions of years ago)

When eon began	Eons	Eras	Periods		Epochs	
4,500	Hadean	Prephanerozoic	beginning and ending dates for periods and epochs (in millions of years ago)			
3,900	Archean					
2,500	Proterozoic					
580	Phanerozoic	Paleozoic 580–245	Cambrian	580–500		
			Ordovician	500–440		
			Silurian	440–400		
			Devonian	400–345		
			Carboniferous	345–290		
			Permian	290–245		
		Mesozoic 245–66	Triassic	245–195		
			Jurassic	195–138		
			Cretaceous	138–66		
		Cenozoic 66–0	Paleogene	66–26	Paleocene	66–54
					Eocene	54–38
					Oligocene	38–26
			Neogene	26–0	Miocene	26–7
					Pleiocene	7–2
					Pleistocene	2–0.1
0					Recent	0.1–Now

* Not to scale and simplified

the evolution of life in the microcosm, as well as of the microcosm and its relentless expansion.

Over the first million years of expansion after the Big Bang, the universe cooled from 100 billion degrees Kelvin, as estimated by physicist Steven Weinberg, to about 3000 degrees

K, the point at which a single electron and proton could join to create hydrogen, the simplest and most abundant element in the universe.[14] Hydrogen coalesced into supernovae—enormous clouds that over billions of years contracted from cosmic to submicrocosmic densities. Under the sheer force of gravity, the cores of the supernovae became so hot that thermonuclear reactions were fired, creating from hydrogen and various disparate subatomic particles all the heavier elements in the universe that we know today. The richness of hydrogen is in our bodies still—we contain more hydrogen atoms than any other kind—primarily in water. Our bodies of hydrogen mirror a universe of hydrogen.

The newly created elements spewed off into space as the dust and gas that compose the galactic nebulae. Within the nebulae, more stars and sometimes their satellite planets were born, again as particles of dust and gas gravitated toward each other, falling in and concentrating until nuclear reactions were generated. Before the first matter that could be called the Earth gathered within our solar nebula at an outer arm of the Milky Way, five to fifteen billion years and billions of coalescing events forming the stars of the universe had already occurred.

In the cloud of gases destined to become Earth were hydrogen, helium, carbon, nitrogen, oxygen, iron, aluminum, gold, uranium, sulfur, phosphorus, and silicon. The other planets in our solar system began as similar clumps of gas and dust particles. But all would have cooled and floated about as the aimless detritus of lifeless space were it not for the huge star that formed from the center of the nebula, pulling the hardening smaller bodies into orbit and igniting into a stable, long-lasting burn that bathed its satellites in continuous emanations of light, gas, and energy.

At this point, about 4,600 million years ago, the Earth mass was already in circumstances that were to suit it for the emergence of life. First, it was near a source of energy: the sun. Second, of the nine major planets orbiting the sun, the earth mass was not close enough so that its elements were all blown away as gases or all liquefied as molten rock. Nor was it far enough away for its gases to be frozen as ice, ammonia, and methane as they are today on Titan, the largest moon of Saturn. Water is liquid on Earth but not on Mercury where it has all evaporated into space or on Jupiter where it is ice. Finally, the Earth was large enough to hold an atmosphere, enabling the fluid cycling of elements, yet not so large that its gravity held an atmosphere too dense to admit light from the sun.

When the sun ignited, an explosive blast of radiation swept through the nascent solar system, stirring up the early atmospheres of the earth and other inner planets. Hydrogen, too light a gas to be held by the earth's gravity, either floated into space or combined with other elements, producing ingredients in the recipe for life. Of the hydrogen that was left, some combined with carbon to make methane (CH_4), some with oxygen to make water (H_2O), some with nitrogen to make ammonia (H_3N), and some with sulfur to make hydrogen sulfide (H_2S).

These gases, rearranged and recombined into long-chained compounds, make practically every component of our bodies. They are still retained as gases in the atmospheres of the massive outer planets, Jupiter, Saturn, Uranus, and Neptune, or as solids frozen into their icy surfaces. On the smaller, new, and molten Earth, however, phenomena more complex than gravity began to involve these gases in cyclical processes that would keep them here to the present day.

The fury and heat in which the early Earth was formed was such that during these first years of the Hadean Eon

(4,500–3,900 million years ago) there was no solid ground, no oceans or lakes, perhaps not even the snow and sleet of northern winters. The planet was a molten lava fireball, burning with heat from the decay of radioactive uranium, thorium, and potassium in its core. The water of the Earth, shooting in steam geysers from the planet's interior, was so hot that it never fell to the surface as rain but remained high in the atmosphere, an uncondensable vapor. The atmosphere was thick with poisonous cyanide and formaldehyde. There was no breathable oxygen, nor any organisms capable of breathing it.

No Earth-rocks have survived this hellish primeval chaos. The Hadean Eon is dated from meteorites and from rocks taken by Apollo astronauts from the airless moon, which began to cool 4,600 million years ago while the Earth was still molten. By about 3,900 million years ago, the Earth's surface had cooled enough to form a thin crust that lay uneasily on the still-molten mantle, the structure below it. The crust was punctured from below and impacted from above. Volcanoes erupted at cracks and rifts, violently spilling their molten glass. Meteorites—some as huge as mountains and more explosive than the combined nuclear warheads of both superpowers—made violent crash landings. They cratered the chaotic terrain, sending up vast plumes of dust which were rich in extraterrestrial materials. The dark dust clouds, swept by vicious winds, swirled around the globe for months before finally settling down. Meanwhile, tremendous frictional activity caused widespread thunderclaps and electrical lightning storms.

Then, 3,900 million years ago, the Archean Eon began. It was to last for one-and-a-third billion years, and was to see everything from the origin of life to its spread as soft, colorful, purple and green mats and hard, rounded domes of bacteria. The immense amounts of rock that 3,000 million years later would become the American, African, and Eurasian land masses floated about the globe in the unfamiliar shapes of ancient continents. The recognizable continents appeared in their present positions in only the last tenth of a percent of our planetary history.

Heat and radioactivity still brewing in the Earth's core sent lava boiling up through cracks in the just-cooling crust. Much of the lava contained molten magnetic iron whose molecules oriented themselves to the Earth's magnetic pole as it froze into rock. In the early 1960s, studies of these ancient magnetic orientations confirmed what earlier eyes had observed from the shapes of continents and the correspondence of rock layers and fossil wildlife at their edges: the several "plates" into which the Earth's crust is split move about on the molten mantle, separating from some and crashing into others as they shift. Moving up to centimeters a year, a continental plate can cover a hundred miles in a million years. Two hundred million years ago, for example, India was attached to Antarctica, far from the rest of Asia. Drifting nearly two inches a year, India moved northward over 4,000 miles, joining the Asian continent only about sixty million years ago.

The seams between the plates host violent activity. Where the plates are separating and magma boils up to fill the widening rifts, new land or ocean floor is created. Where they collide, earthquakes and volcanoes abound and the Earth is thrown up into mountains. The slow but violent confrontation between the Indian and Asian plates thrust Mount Everest and the Himalayas to the peak of the world.

Today the quakes and tremors along the San Andreas fault in California signal the inexorable progress of the huge Pacific plate, moving northwest as it collides against the northward-moving plate from which the North American continent sticks up. And in North Africa, the Zambezi River in Mozambique traces a split in the earth's armor—the Great African rift—that is cracking the continent of Africa apart. Toward the south, huge amounts of water fill the cracks as soon as they form; great volumes of rock are caving in. Toward the north, at Afar in Ethiopia, water has not yet obscured the view. Molten rock oozes toward the surface and freezes into "pillow basalts" to form the floor of a new Pan-African ocean. The floor of this future ocean is still largely dry. And the panoramic view of the Afar valley is just what you would see if the water of the Atlantic Ocean were drained and you could watch the formation of new sea floor along a rift zone.

The San Andreas fault, the African rift, the Mid-Atlantic rift, the East Pacific rise, and the volcanic islands of Hawaii are rare sites of earth-building activity on a largely placid planet today. But during the Archean Eon, the Earth's surface was riddled by such tectonic activities. Huge quantities of steam shot out of blow holes and splitting seams. The Earth lay covered in a darkening fog of carbon gases and sulfurous fumes. Showers of icy comets and carbonaceous meteorites bombarded the planet, burning through the atmosphere to the weak and unstable surface, further rupturing the crust. Carbon and water came with them from space in sufficient quantities to add to the Earth's own supplies of what were later to become the staples of life.

As the Earth's surface continued to cool, the clouds of steam filling the atmosphere could finally condense. Torrential rains fell for perhaps a hundred thousand years without cease, creating hot, shallow oceans. Submerged plate boundaries, rich in chemicals and energy, steadily vented hydrogen-rich gases into the seas. Water hitting the boiling lava in rifts and volcanoes evaporated, condensed, and rained down again. The waters began to erode the rocky landscape, smoothing out the pockmarks and wounds made by the constant belching of volcanoes and powerful impacting of meteorites. The waters rounded off the mountains as they were created, washing minerals and salts into the oceans and land pools. Meanwhile, in an event sometimes called the Big Belch, tectonic activity released gases trapped in the Earth's interior to form a new atmosphere of water vapor, nitrogen, argon, neon, and carbon dioxide. By this time much of the ammonia, methane, and other hydrogen-rich gases of the primary atmosphere had been lost into space. Lightning struck. The sun continued to beam heat and ultraviolet light into the Earth's thickening atmosphere, as the fast-spinning planet spun in cycles of five-hour days and five-hour nights. The moon too had condensed from the sun's nebula. Since some 15 percent of the moon is material of Earth origin, the best recent model suggests that the moon arose when a planetoid crashed into the Earth's surface but could not completely escape Earthly gravity, going into orbit. Our faithful natural satellite, rather large for a puny inner planet like the Earth, from the beginning pulled rhythmically on the great bodies of water, creating tides.

It is from this Archean Eon, from 3,900 to 2,500 million years ago, that we have found the first traces of life.

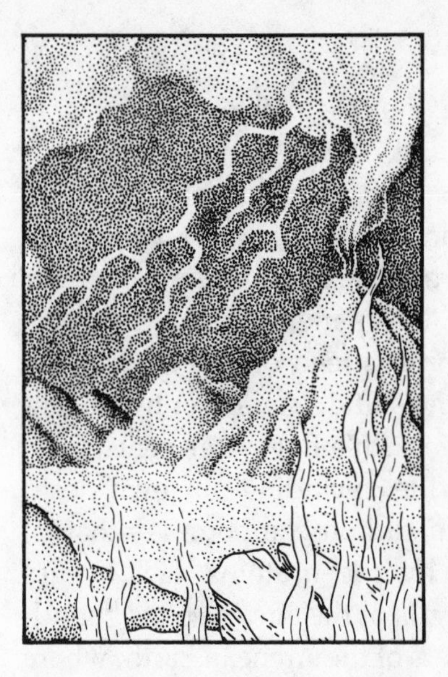

CHAPTER 2

The Animation of Matter

FEW quests are so magical as that for the origins of life. Scientists, alert for any clue, have amassed a telling body of data. A new underwater world, relevant to our thinking about the origins of life, was discovered in 1973. Oceanographer Jack Corliss, a professor at Oregon State University, saw for the first time undersea continental plate seams where magma, steam, and gases still mingle with salty water as they did everywhere in Archean times. Except for an occasional deep-sea fish and the tenacious films of the hardiest microorganisms, the pitch-black, cold (4° centigrade) floor of today's ocean is nearly everywhere barren. Yet along the seams of the earth's great plates where sulfide spews up from the hot mantle below, there are peculiar communities of underwater creatures. At such sites near the Galapagos Islands on the equator, off the shores of Baja California in Mexico, and 3,400 meters beneath the water in the Gulf of Mexico a few hours west of St. Petersburg, Florida, oceanographers have found giant red tubeworms of the genus *Riftia*.

So called because the rifts in the ocean floor are the only place they have ever been found, *Riftia*—as well as various fish, giant clams, other worms, and an occasional octopus—surround the crevices and cracks. None of these animals of the abyss feeds on plants. Plants, algae, any photosynthetic life forms, need light, but no light penetrates to the bottom of the sea. Instead, the rift animals feed on stringy bacteria that derive their energy from the sulfide and other hydrogen-rich gases emanating from the Earth's hot-water vents.

The chemistry of which all life, including our own flesh, is composed is that of reduced carbon compounds—that is, carbon atoms surrounded by hydrogen atoms. Jack Corliss believes that life could have begun at the ancient plate boundaries in the shallow, warm waters of the Archean Earth, where hydrogen-rich gases from the Earth's interior reacted with the carbon-rich gases of the atmosphere. (Indeed, 90 percent of the carbon of our own bodies is estimated to have passed at one time or another through such plate seams and vents.)

The flexibility of carbon is one of the secrets of life on Earth. In their highly agitated states during the hot, wet, and molten Archean conditions, carbon atoms combined rapidly with hydrogen, nitrogen, oxygen, phosphorus, and sulfur to generate a vast diversity of substances. This collection of carbon-containing molecules has continued to exist, interact, and evolve. Those six elements are now the chemical common denominator of all life, accounting for 99 percent of the dry weight of every living thing. Moreover, the percentage of each of these elements, the proportion of amino acids and genetic components, and the distribution of long protein and DNA macromolecules in the cells are similar in all forms of life, from bacteria to the human body.[15] Like Darwin's recognition of the essential similarities of apes and humans, these chemical similarities point to a common ancestor for all life, and furthermore, to the sorts of conditions that must have existed on the early Earth when there was little chemical difference between living cells and their immediate environment.

In 1953, a celebrated series of experiments at the University of Chicago launched a new field of laboratory science, variously referred to as "prebiotic chemistry," "primitive Earth-model experiments," or "experimental chemical evolution." Stanley L. Miller, a graduate student of the Nobel Prize–winning chemist, Harold C. Urey, bombarded a mock-up of the primary atmosphere (a mixture of ammonia, water vapor, hydrogen, and methane) with a lightninglike electrical discharge for a week. He was rewarded with the production of the two amino acids alanine and glycine as well as many other organic substances thought before then only to be produced by living cells. (Small molecules of fewer than a dozen or so carbon, nitrogen, hydrogen, and oxygen atoms, amino acids are the components of all proteins.)

Since the Miller-Urey experiments, almost every simple component of the complex molecules of cells has been produced in the laboratory by subjecting various mixtures of simple gases and mineral solutions to different energy sources—electric spark, silent electric discharge, ultraviolet radiation, and heat. Satisfyingly enough, the four most abundant amino acids in the proteins of all organisms are the most easily formed. The indispensable compound adenosine triphosphate (ATP), a molecule that stores energy inside all cells, and other triphosphate precursors to the nucleotides (the structural bases of genes) also can be formed in these sorts of experiments. In some of the most recent studies, all five of the nucleotide parts that compose DNA and its partner molecule RNA—adenine, cytosine, guanine, thymine, and uracil—could be found in mixtures after methane, nitrogen, hydrogen, and water gases had been bombarded with electric sparks. RNA (ribonucleic acid), like DNA, is a long molecule needed for the functioning and reproducing of every single cell of every living being. RNA too carries information and is made of nucleotide bases, sugars, phosphoric acid—all molecules that could have formed by solar radiation on the Hadean earth.

Like the early Earth's environment, these experimental concoctions yield all sorts of other suggestive organic compounds. Their identity and possible function are often a mystery to the human investigators—but not at all to passing microbes. If not protected under sterile conditions, and as long as they are in water solution, the most complex lab-synthesized chemical compounds are quickly gobbled up by modern airborne bacteria and fungi. The microbes, which are nearly everywhere, simply land in the water for a quick meal.

Although no cells have yet crawled out of a test tube, chemist Leslie Orgel of the Salk Institute has recently discovered a 50-nucleotide-long, DNA-like molecule that formed spontaneously from simple carbon compounds and lead salts in the total absence of living cells or complex compounds. Professor Manfred Eigen and his coworkers at the Göttingen Institute in Germany have made short RNA molecules that have replicated by themselves in the total absence of living cells. The late Sol Spiegelman and his colleague Donald Mills of Columbia University synthesized in the test tube infective viruses of the RNA variety that are fully capable of continued replication inside their bacterial hosts. (Unable to create all the components needed to be a true living system, viruses are little more than a stretch of DNA or RNA coated with protein.) Although Spiegelman used only an enzyme (a reusable biochemical that accelerates chemical reactions), a nucleic acid (RNA in this case), and the small molecular precursors of nucleic acids called nucleotides, he employed a form of energy that did not exist on the early Earth: human effort and dollars.

Such experiments have popularized the belief among scientists and lay persons alike that one or a few lightning bolts striking the rich, chemical "soup" of the Archean oceans could have just happened to fuse carbon and hydrogen atoms together with other elements in the right combinations to produce life. One common concept is that life appeared suddenly and almost instantaneously from the prebiotic soup. Other scientists argue that the odds against such instant life are beyond the astronomical—more unlikely than the assembly of a Boeing 707 by a hurricane in a junkyard. But there is no credible way to assess the probability of life's spontaneous generation from nonlife. As origins-of-life expert Leslie Orgel quipped, "We don't even know within a factor of 10^{20} if other soups are viable." All we know is that life did arise. Some theorists feel forced to postulate that the Earth must have been seeded by meteorites carrying the finished molecules of life. They point to the fact that compounds related to the five different kinds of nucleotides as well as amino acids have been found in meteorites.

Yet the conclusions of both "instant life" and "life from meteorites" veer away from the crucial point: that the proper milieu for the slow brewing of early life from nonlife was the early Earth. There was sufficient time and energy available for life's molecular combinations to arise from chemical alliances encouraged by the cyclically changing, energy-charged environment. Besides, chemicals do not combine randomly, but in ordered, patterned ways. There is no need to postulate the unlikely when evidence for the likely abounds. The presence of organic compounds in meteorites only seems to confirm that a hydrogen-rich environment exposed to energy in the presence of carbon—conditions that certainly existed throughout our solar system, if not the universe—will, by the rules of chemistry, produce the building blocks of life. It is the many other unique qualities of the Earth, including its wetness, balmy temperatures, and gravitational properties, that made it a better environment for these molecules than the other planets. The Earth's conditions favored certain chemical combinations more than others, and over time a direction was set.

The ponds, lakes, and warm, shallow seas of the early Earth, exposed as they were to cycles of heat and cold, ultraviolet light and darkness, evaporation and rain, harbored their chemical ingredients through the gamut of energy states. Combinations of molecules formed, broke up, and reformed, their molecular links forged by the constant energy input of sunlight. As the Earth's various microenvironments settled into more stable states, more complex molecule chains formed, and remained intact for longer periods. By connecting to itself five times, for example, hydrogen cyanide (HCN), a molecule created in interstellar space and a deadly poison to modern oxygen-breathing life, becomes adenine ($H_5C_5N_5$), the main part of one of the universal nucleotides which make up DNA, RNA, and ATP.

With no oxygen in the atmosphere to react with and destroy them, amino acids, nucleotides, and simple sugars could form and remain in solution together. Even ATP, a molecule used by all living cells without exception as a carrier for energy, could form from the union of adenine with ribose (a sugar with five carbon atoms) and three phosphate groups.

Some molecules turned out to be catalysts: they made it easier and faster for other molecules to join or split without themselves being destroyed. Catalysts were important before life because they worked against randomness to produce order and pattern in chemical processes. Gradually, they and the reactions they facilitated proliferated more than other combinations. Although increasingly complex, these processes had lasting power. They endured in the waters of the early Earth. Today, certain groups of molecules can self-catalyze a series of surprisingly intricate and orderly or cyclical reactions, each change bringing about another in the molecular chain. Some of these "dead" autocatalytic reactions form patterns whose increasing complexity over time is reminiscent of life.

From both theoretical calculations and laboratory evidence, it has been suggested that an interaction of two or more autocatalytic cycles could have produced a "hypercycle." Some scientists theorize that such catalyzing compounds "competed" for elements in the environments, thus automatically limiting their existence. But the basic idea of the hypercycle is quite the opposite. Far from destroying each other in a fight for chemical survival, self-organizing compounds complemented each other to produce lifelike, ultimately replicating, structures. These cyclical processes formed the basis not only of the first cells but of all the myriad structures based on cells and their products that followed. Cyclical processes are very important to life. They allow life to preserve key elements of its past despite the fluctuations and tendency toward disorder of the larger environment.

The more protected and concentrated the chemicals were, the longer and more complex and self-reinforcing their activities could become. Some may have been shielded inside bubbles or held on the regular surfaces of clays and crystals. Nature's Archean experiments with long hydrocarbon chains were yielding compounds that could encapsulate a droplet of the surrounding water and its contents yet allow movement of other chemicals in and out of the enclosure. This was the semipermeable membrane, a sort of soft door that permitted the entry of some chemicals while prohibiting that of others. The chemical components came together to form membranes and, in the business of making life anyway, membranes are marvels of simplicity. Indeed, the events that led up to their formation have been duplicated in the laboratory under conditions of temperature, acidity, and cycles of wetness and evaporation that are common on the Earth.

A hydrocarbon chain linked to a group of phosphorus and oxygen atoms manifests an electrical charge on the end bearing the phosphate group and no charge on the other end. The chemical as a whole attracts water on its charged end and repels it on the noncharged end. Such chemicals, called phospholipids, tend to line up side by side with each other, the noncharged ends pointing away from the water while the charged ends point down into it. (This is essentially what happens when a drop of oil enters water, instantly forming a film.) These and other types of lipids tend spontaneously to fold into drops, secluding materials on the inside from those on the outside. They have also been shown to form double layers when waves bring two water surfaces, filmed with lipids, together. When this happens, the charged ends of the sheet of lipid molecules point toward each other, sandwiched between the noncharged ends. In this way, the first membranes were formed—the first semipermeable boundaries between "inside" and "outside"; the first distinction between self and nonself.

The membranes of today's organisms are composed of several different kinds of lipids, proteins, and carbohydrates, their functions so complex and precisely calibrated that we are far from fully understanding them all. But the first phospholipid membrane, unlike various other encapsulating structures that can also form in nature's crucible, could, by virtue of its chemical properties alone, concentrate a solution of other carbon chemicals. It could keep potentially interacting components in close proximity; permitting "nutrients" to enter while preventing water from escaping. The membrane makes possible that discrete unit of the microcosm, the bacterial cell. Most scientists feel that lipids combined with proteins to make translucent packages of lifelike matter before the beginning of life itself. No life without a membrane of some kind is known.

There is still a missing link between the most complex concoctions of the working scientist and the simplest viable cell, both in theory and in the laboratory. The gap between small organic chemicals, such as amino acids and nucleotides, and larger biochemicals, such as RNA and protein, is enormous. But a few hundred million years of molecular activity is a long, long time. Scientists have been working only a few decades to provide conditions conducive to the origin of laboratory life, and have come very far. It is not inconceivable that before the turn of the twentieth century a live cell will be spontaneously generated in the laboratory. Given millions of years, the chances of spontaneously forming hypercycles were immensely greater than those available to research workers, who must substitute planning for the blind perseverance of time if they are ever to recreate life.

Probably not once, but several times, amino acids, nucleotides, simple sugars, phosphates, and their derivatives formed and complexified with energy from the sun within the protection of a lipid bubble, absorbing ATP and other carbon-nitrogen compounds from the outside as "food." Fairly complex structures have formed spontaneously from lipid mixtures in the laboratory. For example, David Deamer at the University of California at Davis has observed that some nucleotides are taken up and surrounded by spheres of lipid if the proper ingredients are mixed under appropriate conditions. Bubbles of lipids split in two at first simply from the strain of surface tension, each half carrying on its internal activity. Later, the catalyzing molecules within may have begun to actively maintain the lipid membranes. Perhaps when the supply of available component elements in their tiny local niche was exhausted, the protocells simply broke down and disappeared, while others formed in some other tidal pool, each with a slightly different modus operandi.

To be alive, an entity must first be *autopoietic*—that is, it must actively maintain itself against the mischief of the world.[16] Life responds to disturbance, using matter and energy to stay intact. An organism constantly exchanges its parts, replacing its component chemicals without ever losing its identity. This modulating, "holistic" phenomenon of autopoiesis, of active self-maintenance, is at the basis of all known life; all cells react to external perturbations in order to preserve key aspects of their identity within their boundaries. If the external threats are major, normal cyclical processes may be disrupted and schismogenesis may result. *Schismogenesis,* a word coined by the biologist and philosopher Gregory Bateson, refers to cycles in living systems that oscillate uncontrollably. Bateson believed that schizophrenia could be traced to a special kind of schismogenesis, in this case an overabundance of feedback in the brain leading to mental disintegration. But this is only one highly specific example of the failure of normal cyclical processes. In organisms such as plants and animals, we recognize autopoiesis generally as "health." Schismogenesis is its opposite. But even the predecessors to cells must have had some sort of autopoiesis, the ability to maintain their structural and biochemical integrity in the face of environmental threats.

Once able to stay itself, a structure on its way to becoming living must reproduce itself. Before cells, life and nonlife may have been indistinguishable. The first cell-like systems were what the Belgian Nobel Prize–winning physicist Ilya Prigogine has termed "dissipative structures"—objects or processes that organize themselves and spontaneously change their form. With an influx of energy, dissipative structures may become more instead of less ordered. The sort of information theory that has been so useful in communication technology applies solely to information which consists almost entirely of confirmation. In dissipative structures, information begins to organize itself; pockets of elaboration arise.

From dissipative structures and hypercycles emerged the chain of nucleotides, ribose, and phosphate that can both replicate itself and catalyze chemical reactions. This chain is ribonucleic acid, or RNA, the first sentence in the language of nature. Not yet autopoietic, but highly structured, early RNA in spheres surrounded by strings of lipids accumulated in warm, organically rich waters on a benign Earth. With no predators and plenty of energy, complexification followed. On the Hadean Earth before the dawn of life two chemical trends took hold: self-reference and autocatalysis. Chemicals reacted cyclically, producing versions and variations of themselves that tended to create an environment favorable to the repetition of the original reactions. Autopoietic structures took organization a step further: they used energy to actively and successfully maintain themselves in the face of serious external perturbations. Their boundaries became distinct. This gave them both identity and memory. Today, although all of the chemicals in our bodies are continually replaced, we do not change our names or think of ourselves as different because of it. Our organization is preserved, or rather it preserves itself. From dissipative structures to RNA hypercycles to autopoietic systems to the first crudely replicating beings, we begin to see the winding road that self-organizing structures traveled on their journey toward the living cell.

CHAPTER 3

The Language of Nature

ACCORDING to the *Book of Genesis*, God halted construction of a majestically high tower in Shinar by introducing many languages. The Tower of Babel never reached heaven because its builders, stripped of their common tongue, became confused. This parable shows the importance of a universal language. While people still speak in many languages (though fewer as time goes by), the genetic code—the translation of genes into proteins—is everywhere the same.

Learning the now universal RNA/DNA-based genetic language that emerged from the babel of Archean chemicals made the new science of molecular biology thrilling indeed in the past two decades. The genetic code is a unique phenomenon. The DNA or RNA molecule can replicate itself exactly; but it can also cause the uniform assembly of those other long biochemicals, proteins. This was the central insight of the molecular biological revolution that began when James Watson and Francis Crick discovered the structure of DNA in 1953.

As miraculous as it seems, replication is, on a molecule-to-molecule basis, a shockingly straightforward chemical process. A complementary chemical structure prescribes the shape and properties of replicating molecules: DNA and RNA are one lengthwise half of a single long molecule. Like the matching teeth of a zipper, in the presence of the right ingredients, the components of the missing half simply line up and fit.

RNA is a particularly versatile sort of half-molecule. It can match up another long RNA like itself, or it can match up short bits of nucleotides with amino acids attached—producing all the proteins that give organisms their varied shape and forms. RNA's components are the four different nucleotides, the bases adenine, guanine, cytosine, and uracil, each of which holds onto a phosphate group (composed of phosphorus and oxygen) and ribose, a kind of sugar. Taken in groups of three, the sequence of nucleotides in one type of RNA can be a signal for a second type of matching RNA to attach to amino acids in its environment. As the amino acids link up, one after another, a protein is formed, and this protein can, in turn, accelerate the further matching of the RNA molecule, thereby producing more RNA.

The first membrane-enclosed autopoietic bodies were probably governed only by RNA. They could replicate themselves by making proteins that made more RNA. The development of the double-stranded, far longer and less accident-prone DNA molecule probably came later, gradually taking on the function of a mold or template for the copying of RNA.

DNA, too, is made of only four nucleotides, each with a sugar, and a phosphate group. DNA has thymine instead of RNA's uracil and its sugar is deoxyribose instead of ribose. The two intertwined DNA and RNA strands fit together with adenine always linking with thymine and guanine always with cytosine, again because of their chemical structure. The smallest bacterium has hundreds of thousands of these paired parts, the so-called base pairs; animal and plant cells many millions. All cells today have both DNA and RNA. The line-up of nucleotides leads to the line-up of proteins, which makes more nucleotides and makes them line up. This arrangement of chemicals is not only dissipative and autopoietic, it is the reproducing ancestor of every life form on Earth.

Proteins make an organism what it is. The line-up of nucleotides specifies the composition and quantity of proteins. Organisms differ largely because the sequence of spiraling nucleotides in DNA molecules differs. Variations in the order and number of nucleotide pairs lead to the fabrication of different proteins. At least several thousand different proteins in each cell determine how the organism looks, how it moves, how its metabolism runs. Every cell needs proteins to speed up chemical reactions. Without certain proteins, many essential biological reactions would take place very slowly or grind to a halt. Chaos and ordinary chemistry would reign.

The common denominator of life extends further. Only about twenty different amino acids, linked in chains of a few dozen to several hundred, make up the proteins in all known organisms on Earth. The amino acid sequence, primarily, determines the protein's shape, and the shape determines its function. The code for translating the sequence of nucleotides in DNA to a sequence of amino acids in a protein is nearly universal. In almost all cases, a given nucleotide sequence will translate into the same amino acid sequence.

In all organisms, each triplet of nucleotides on the coding nucleic acid—called a codon—specifies one amino acid. But there are signs that two-nucleotide codons may have constituted an early version of this system. For example, the third nucleotide in a codon is often redundant: uracil, adenine, cytosine, or guanine could each be the third after a cytosine-guanine doublet to form a triplet on messenger RNA. In any of the four cases the amino acid arginine would be coded for. Additionally, the middle nucleotide often determines the simplest and most common amino acids. The early genetic code was no doubt simpler and less faithful than it is today. A living language, it still carries evidence of its etymological roots.

Like words, too, the elements of the code can be tampered with, rearranged, changed, and passed down in an altered form. Mutations are heritable changes in the quantity or sequence of DNA bases. A mutation occurs when something in the environment—radiation, say—either breaks a chemical bond or forges an uncalled-for one, and the resulting change in the DNA sequence, which confers new abilities or disabilities, is copied and passed down through the cell's descendants or causes the cell's demise. Like the difference of an "s" between the words *laughter* and *slaughter*, small changes or additions can have synergetic effects.

In the excitement following the discovery of DNA's and RNA's vital role, the lion's share of responsibility for life's diversity was attributed to minute, base-pair mutations in these molecules. But at an estimated rate of one base-pair mutation per group of a million to a billion cells in each generation, even base-pair mutations seem inadequate to explain life's grand variety of organisms.

In language, the swiftest and most telling changes come about through usage. Street argot and slang, produced by usage at its most basic, everyday level, constantly filter into the mainstream of language and eventually find their way into official dictionaries. As we shall discover in the next chapter, the "street" where the genetic message undergoes continual and rapid changes is the microcosm.

The reading and copying of the genetic message can happen in an incredibly short time. In seconds to minutes, proteins are assembled from amino acids. Although isolated DNA cannot replicate, if placed in a test tube with proteins (catalysts), nucleotides (food), and an energy supply (just the chemical energy in the nucleotides), DNA can make a copy of itself within seconds. To the great possible benefit of future medicine, it can even do so after having been frozen in solutions in glass vials for several years. Although not capable, as are whole organisms, of self-maintenance, DNA in the proper chemical milieu can replicate.

In 1977, Sir Frederick Sanger and his fellow researchers at a medical research laboratory in Cambridge, England, decoded the first complete genetic message and uncovered a new twist in the language. The DNA of a virus called ϕX174 is only 5,375 nucleotides long, which would account for approximately 1,792 amino acids—good for about five proteins—yet it codes for nine different kinds of protein, which require about 3,200 amino acids. Sanger's group discovered that using the same DNA one stretch of nucleic acid determines more than a single protein, depending on where the message-reading begins. This seems a startling molecular "invention." Yet, once again considering that life keeps itself together at all costs, using nucleotide messages to specify proteins, it doesn't seem so unlikely that the genetic code should have developed double entendres. Indeed, a single length of nucleotide is read for different meanings in several kinds of cells and in some mitochondria as well. There is ambiguity in the universal language of life.

In the study of complex animals and plants, traits such as lungs or eyes or flowers that require the interaction of many genetic factors are called *semes*. First life, too, evolved because of conditions favoring not individual traits—this enzyme or that nucleotide base pair—but semes, chains of chemical reactions that yielded food or movement or some other critical aptitude. In microbes semes tended to be metabolic. For instance, microbes that made the metabolites they needed by using carbon dioxide from the air starved far less often than those unable to feed from carbon in the air. The biochemical steps allowing the use of atmospheric carbon dioxide is an example of a microbial seme.

It seems silly to postulate a single dramatic moment of magical lightning when DNA and RNA spontaneously formed a cell and life began. Many dissipative structures, long chains of different chemical reactions, must have evolved, reacted, and broken down before the elegant double helix of our ultimate ancestor formed and replicated with high fidelity. Indeed, living forms based on totally different types of replicating molecules may have arisen and developed for a while before disappearing altogether. But because they are the common denominator of all life today, it is clear that at some point lipid membranes containing RNA and DNA began to flourish. The numbers of these tiny bacterial spheres increased and diminished in a process of ebb and flow. To borrow Julian Huxley's simile, the waves break and go back on themselves, yet the tide rises just the same. At some point some time before 3,500 million years ago, the evolutionary tide reached the level of life as we know it: that of the membrane-bounded, 5,000-protein, RNA-messaged, DNA-governed cell. Once autopoiesis ensured its existence and reproduction guaranteed its expansion, evolution was under way. The Earth's microcosm, the Age of Bacteria, had begun.

The tiny Archean sacs of DNA and RNA carried out their activities prodigiously. With sleep unknown to them they grew, consumed energy and organic chemicals, and divided incessantly. Their colonies and fibers interconnected and covered the sterile globe in a spotty film. The dimensions of this film have expanded into a patina of life, or biosphere, the place where life exists. Today the biosphere surrounds the Earth from a little deeper than six miles into the ocean to over seven miles up, above the mountain tops at the top of the lower atmosphere, called the troposphere. Bacteria first expanded in the waters, where they modified the liquid and produced gases. They then expanded to the surfaces of the sediments, where they still survive. None lived its complete life in the atmosphere, nor can any being do so today. Nonetheless, mostly in the form of dormant particles (such as seeds, spores, and eggs), some organisms can be found spending some time in the atmosphere. The biota—all life on Earth—becomes tenuous at its extremes several miles up and down. The core of life on earth, the dense, thriving center of the biota has been and still is within a few meters of the Earth's surface. Dr. Sherwood Chang of NASA Ames Research Center has suggested that life probably began at the interface of liquid, solid, and gaseous surfaces where there is an energy flux and dissipative structures can easily form. Today life still thrives where water meets the land and air. The biota, the sum of all life, primarily as the microbiota, the sum of all microbial life, is ancient, extending through the vast biosphere. Over time it has spread out. Yet from the point of view of chemical and metabolic innovation, the biota at its core has not significantly changed. Composed of all reproducing beings, continuous through time, the planetary patina has a life of its own. The biota cycles inorganic substances, such as rocks, muds, and gases, modulating and controlling itself. Cells collectively preserve the water-, carbon-, and hydrogen-rich conditions of their origin. The biosphere retains in its midst gases such as hydrogen and methane that otherwise would long ago have been lost from the Earth by cosmic processes. It is a souvenir of itself.

In a sense, the essence of living is a sort of memory, the physical preservation of the past in the present. By reproducing, life forms bind the past and record messages for the future. The oxygen-shunning bacteria of today tell us about the oxygenless world in which they arose. Fossil fish tell us of open bodies of water in continuous existence for a hundred million years. Seeds that require freezing temperatures to germinate tell us of frozen winters. Our own human embryos represent phases of animal history in their stages of development.

Put another way, life is extremely conservative. On whatever level—the individual organism, the species, the biota as a whole—life expends energy such that it preserves its past, even if, paradoxically, various threats force it to innovate. Since autopoiesis is an imperative of the biota as a whole, life will expend huge quantities of energy to preserve itself. It will change in order to stay the same.

There is little doubt that the planetary patina—including ourselves—is autopoietic. Life at the surface of the Earth seems to regulate itself in the face of external perturbation, and does so without regard for the individuals and species that compose it. More than 99.99 percent of the species that have ever existed have become extinct, but the planetary patina, with its army of cells, has continued for more than three billion years. And the basis of the patina, past, present, and future, is the microcosm—trillions of communicating, evolving microbes. The visible world is a late-arriving, overgrown portion of the microcosm, and it functions only because of its well-developed connection with the microcosm's activities. Microbes by themselves are thought to have maintained the mean temperature of the early Earth so that it was hospitable for life, despite the much cooler "start-up" sun that astronomers believe existed then. During Archean times "stupid" microbes also continually modified the chemical composition of the atmosphere so that it did not become prohibitive to life as a whole. We know from the continuous fossil record of life that the temperature and atmosphere of the planet never destroyed all life. Barring divine intervention and luck, only life itself seems powerful enough to have promoted the conditions favoring its own prolonged survival in the face of environmental adversity.

Grasping as best we can the formidable powers of the biosphere in which we live out our lives, it is difficult to retain the delusion that without our help nature is helpless. As important as all our activities seem to us, our own role in evolution is transient and expendable in the context of the rich layer of interliving beings forming the planet's surface. We may pollute the air and waters for our grandchildren and hasten our own demise, but this will exert no effect on the continuation of the microcosm. Our own bodies are composed of one thousand billion (10^{12}) animal cells and another *ten* thousand billion (10^{13}) bacterial cells. We have no natural "enemies" that eat us. But after we die we return to our forgotten stomping ground. The life forms that recycle the substances of our bodies are primarily bacteria. The microcosm is still evolving around us and within us. You could even say, as we shall see, that the microcosm is evolving *as* us.

CHAPTER 4

Entering the Microcosm

IN 1977 on the periphery of a tiny South African mountain town called Fig Tree, Elso S. Barghoorn, a paleontologist at Harvard University, hacked out pieces of flintlike rock from the side of a worn mountain in the Barberton Mountain Land and collected samples. Back in Cambridge, Massachusetts, he cut the rock into slices so thin that light could be seen through them. He placed the thin rock samples under a microscope.

Water loaded with minerals from nearby volcanoes had formed the black chert, which Barghoorn knew from experience is the kind of rock most likely to contain fossils. More than three billion years ago at this place, silica-rich lava repeatedly poured into thick black mud, hardening it into chert. The restless volcanoes spewed ash into the air, which fell in immense piles on the mud and into the water of a now-vanished Swaziland sea that for millions of years covered most of what is now southern Africa. The years of volcanism and erosion, of transport of rubble and stones, piled up many complex layers of rocks. Over vast periods of time these layers covered the shores and lined the floor of the ancient sea. Today records of the bygone scene extend as hills and rock ledges for hundreds of miles in South Africa and the country of Swaziland. In some places the complex layers of Swaziland rock are more than ten miles thick.

The Swartkoppie zone of this fossil ocean is laced by seams of coal-like carbon deposits several hundred meters thick that might be mistaken for the remains of a tropical swamp of trees, seed ferns, and club mosses, like the ones that produced coal in Pennsylvania 300 million years ago. Such carbon-rich deposits in the earth have always meant photosynthetic life. But these Swaziland rocks were laid down 3,400 million years ago—they are more than ten times as old as the swamp forests. (The very first fossil land plants are about 450 million years old.)

Professor Barghoorn had been seeking the antiquity of life. After much study of the thin sections of African cherts in collaboration with his students, Barghoorn discovered hundreds of round objects, most of them simple spheres. But one or two dumbbell shapes caught Barghoorn's attention. Were these life itself—caught in the act of dividing? In other samples from the nearby Kromberg Formation of rocks, thin microscopic filaments were found similar to today's cyanobacteria (blue-green algae). Here were the oldest fossils on the planet—hard evidence that bacteria, already accomplished in photosynthesis, thrived on the Earth only 500 million years after the earth's first rocks had formed.[17]

Barghoorn's South African find was the fruit of a conscious search. More than twenty years earlier, at the University of Wisconsin, the geologist Stanley Tyler had showed him rocks from the northern shores of Lake Superior. The 2,000 million-year-old rocks were full of strange objects that had looked like fossils of microscopic life to Barghoorn. Since 1954 then,

Barghoorn had been on the lookout for early life. Barghoorn's dogged persistence to find the oldest fossils in ordinary-looking rocks led to a thirty-year burst of microfossil research that is still under way.

Until the 1950s it was thought that life began shortly before 570 million years ago, since an explosion of hard-shelled animal fossils, often called the Cambrian explosion, appears in rocks of that age all over the Earth. No clear-cut examples of skeletalized animal fossils in older rocks had ever been found. Forgetting that simpler, softer-bodied animals might not have been preserved, some scientists had assumed a rather sudden appearance not only of all animals but even of life itself.

It turned out that the rocks of England and Wales which held some of the best-studied early animal fossil deposits were missing their late pre-Cambrian layers. When more continuous rock formations were found in China, South Australia, Siberia, and elsewhere, they revealed many excellent sandstone impressions of recognizable, soft-bodied marine animals. More recently the pre-Cambrian has been carefully combed by Barghoorn and others to reveal its nonobvious fossils, and new evidence has pushed backward the probable origin of life on the earth.

All the good evidence for early life is not in fossils of the bodies of the organisms themselves. In what is known as the Isua Formation in eastern Labrador and southwestern Greenland, on the outskirts of the polar cap, are found the oldest sedimentary rocks on Earth. These rocks may be the graveyards of what were flourishing films and scums of Archean bacteria. Nearly four billion years old, 3,800 million to be precise, the rocks have been subjected to heat and pressure so enormous that no fossil could possibly remain intact. But life may have left its traces in Isua. Carbon, the central element of life, is abundant in some Isua rocks, and this carbon is found in ratios of carbon isotopes (C^{13} to C^{12}) characteristic of photosynthetic organisms.[18] Was this increase of carbon^{12}, this skewed ratio of the two different forms of carbon the result of bacterial photosynthesis? Does the carbon-rich deposit represent the aftermath of bacterial cell walls, genes, and proteins? The carbon of the Isua rocks is in the form of graphite, produced when shale—a kind of mud turned to stone—is put under intense pressure and heat. If this graphite came from the remains of mud-dwelling photosynthetic bacteria, the date for the origins of life would be pushed to a point almost contemporaneous with that of the Earth's surface.

Barghoorn's Swaziland discovery of actual 3,400 million-year-old fossil microbes raises a startling point: the transition from inanimate matter to bacteria took less time than the transition from bacteria to large, familiar organisms. Life has been a companion of the Earth from shortly after the planet's inception. Indeed, the vital bond between the environment of the Earth and the organisms upon it makes it virtually impossible even for biologists to give a concise definition of the difference between living and nonliving substance.

Since it had to deal with change, as life originated it was forced to maintain itself. The same energetic forces that made crucial chemical bonds tended to break them apart again. The first cells not only had to trap energy to maintain their integrity against any hostile forces; they also needed water and food in the form of carbon-hydrogen-nitrogen compounds. The environment continually changed around them. Whereas reproduction had to originate only once, different adaptive strategies took many complicated steps to achieve. Each meteorite impact, volcanic gas emission, drought, and flood was a crisis. Early life either maintained its integrity—using up the energy and carbon sources in its surroundings—or fell by the wayside.

The challenging work of tracing the development of life's survival strategies in the fossil record has fascinated scientists for the past three decades. Sophisticated techniques of analysis have helped. Translucent slices of microcrystalline quartz, thinner than a piece of paper, are looked at with light, fluorescent, and polarizing microscopes. Under the best conditions the fossilized microbes show up so clearly that only an expert can distinguish them from living bacteria. For example, one fossil, the 2,200 million-year-old photosynthetic bacterium *Eosynochococcus* taken from Belcher Island in Hudson Bay, Canada, is so like the modern bacterium *Synochococcus*, scraped from a Norwegian rock, that distinguishing the two poses considerable difficulties.

There are other ways of studying ancient rocks besides a direct examination of the fossils they contain. Most often fossils, embedded in opaque shales, do not preserve very well, but the organic matter making up their bodies still remains trapped. Sometimes this organic matter can be released by grinding the rocks and treating the fragments with strong solutions. The resulting "rock juice" is analyzed by gas chromatography and mass spectroscopy. These techniques have led to the idea of "chemical fossils." Living organisms of course contain high concentrations of certain chemicals. They also often tend to favor chemical arrangements that are different from those that occur spontaneously in meteorites or prebiotic experiments using the same elements. The presence of certain kinds of carbon chemicals and carbon isotope ratios are taken as clues to ancient life.

Both the search for and the analysis of these rare and enigmatic clues in the rocks are aided by our increasing knowledge of cell biology. Knowing how a cell works tells us what to look for. In tracing life's skills and systems to their beginnings in the fossil record, biologists and paleontologists have launched an all-out search into the mysteries of life's microbial past. We are uncovering an amazing and humbling story of life's accomplishments when it was still in its infancy.

First life—spheres of DNA, RNA, enzymes, and proteins a millionth of a meter in diameter—probably resembled the minimal organisms that exist today. Like first life, minimal life seems to contain DNA with limited ability to direct metabolism. There are not enough genes to take care of making all the amino acids, nucleotides, vitamins, and enzymes needed. Many of the best-studied of these minimal kinds of bacteria are parasites that get what they need from the animals in which they live. In the days when life could absorb its components directly from the environment, it literally freeloaded, using the biochemicals that had accumulated due to exposure of chemical mixtures to ultraviolet light and lightning in the absence of oxygen.

But the chemical feast didn't last. In each little niche, such free nutrients were quickly depleted as the microbes ceaselessly imbibed, grew, and divided. No doubt in the first few million years of life's tenure, each "famine," change of climate, or accumulation of pollution from the microbes' own waste gases always extinguished some and probably sometimes almost all the patches of life on the face of the Earth. Life might have fluctuated in tenuous balance with the rate at which the sun could create more nutrients, or might have quickly died out altogether, were it not for a vital trait: the ability of DNA to duplicate itself, thus leading to extra copies that could playfully, experimentally change.

Replication of DNA is necessary for the continuity of life, but it is not enough for the evolutionary processes. Mutation is absolutely required for Darwin's "descent with modification." By virtue of their tininess and enormous numbers, microbes respond relatively readily to quite major environmental changes. They reproduce without hesitation if food and energy are around. Fast bacteria can divide every twenty minutes or so, in principle yielding in two days 2^{144} individuals.[19] This number is vastly more than the number of human beings who have ever lived. In four days of unlimited growth there would be 2^{288} bacteria. Indeed, this number is greater than the number of either protons (roughly 2^{266}) or quarks that physicists have estimated to exist in the universe, and serves only to remind the reader of the nature of exponential reproduction.

About once in a million divisions, an offspring appears which is not identical to its parent. (Bacteria, since they reproduce asexually by simple division, have only a single parent.) Most mutants are worse off than the parent and die. But a single successful bacterial mutant can quickly expand throughout its environment.

Ordinary hazards of the environment—temperature variations, the quality and quantity of sunlight, the concentration of salts in the water—all acted to diversify the populations of microbes in different places. In the face of starvation, there began to accumulate in the planetary patina a variety of successful new bacteria with new semes—new metabolic pathways by which food and energy could be extracted from various raw materials.

One of the first innovations enabled cells to use sugars and to convert sugar to ATP energy. The lower energy byproducts of sugar—alcohols and acids—were then excreted as waste. DNA absorbs the ultraviolet rays of sunlight. Since sunlight tended to break apart the DNA of bacteria, those bacteria which could afford to stayed in the mud and water away from the surface sunlight. These cells lived off chemicals in the earth; they developed various sugar breakdown processes, known as fermentation, which are still in operation, unchanged, today. Cells can begin with sugars (glucose, sucrose) or with carbohydrates (which are more complex chains of sugars, such as cellulose and starch). Fermenters can also begin with simple nitrogen-containing compounds, such as amino acids, or with alcohols and acids. They can end up with carbon dioxide and ethanol, like the bacteria that ferment the ingredients for wine, beer, and liquor; with lactic acid, like those that sour milk and ripen some cheeses; or with acetic acid and ethanol, like those that form in sewage or make vinegar. The fermentation process generally gives the cell a few molecules of ATP for every small food molecule broken down.

Since the waste products of fermenting bacteria—certain acids and alcohols—still contain energy, fermentation isn't entirely efficient. In time, other microbes evolved that ate the wastes of fermenting bacteria. These new bacteria broke down the wastes, deriving more carbon and energy from them. Bacteria thriving on the waste of other bacteria are much like today's champion excrement eaters: the dung beetles and dung fungi. Such fermentation food chains still exist in swamp and lake muds as well as in tidal flats near the sea. They exist in animal guts and standing puddles, indeed anywhere there is something to ferment and where the quantity of oxygen and light are low. One fermenter's food was another's waste. Fermenters transforming food into waste and complementary fermenters transforming waste into food established a cyclical process of energy-releasing carbon transformations. Fermentation was never lost. When we exert ourselves, for example, by running up the stairs, our cells momentarily bypass their typical oxygen-mediated metabolism and revert to the ancient fermenting mode. Although less efficient a means of deriving ATP, this fermentation metabolism is still with us.

One of the major fermenters that most resembled early

life—a form of bacteria known as clostridia—also evolved the vital function of taking nitrogen gas from the atmosphere and converting it to the ammonialike side chain of amino acids, nucleotides, and other organic compounds. To fix nitrogen takes a prodigious amount of energy—from six to eighteen molecules of ATP for every molecule of nitrogen. To do this industrially (e.g., to manufacture plant fertilizer) requires at least 300 times normal atmospheric pressure at 500 degrees C. No plant or animal is capable of this feat; in fact, neither are most microbes. All organisms depend for their existence on nitrogen-fixing bacteria. Moreover, without this activity of capturing nitrogen gas directly from the air, the Earth's life would long ago have died out because of nitrogen starvation. Far from being included in the proteins of all living cells as it is today, nitrogen would have become inaccessible, trapped as an inert gas in the air. Fermentation releases small quantities of nitrogen from within cells into the air even today; fortunately, nitrogen fixers return it to living organisms. If the world of life had not retained the bacterial trick of nitrogen fixation, we would all have perished of nitrogen deficiency. The clostridia, azotobacters, rhizobia, and other such bacteria have continued to supply the entire biosphere with its vital nitrogen compounds, thus preventing irreversible malnutrition on a global scale.

Of course, the now burgeoning populations of fermenters were still dependent on environmentally produced compounds for food—although a few could use carbon dioxide from the air as well. As the prebiotically produced organic compounds were gobbled up and became scarce, however, the pressures intensified. Another metabolic pathway arose in the ancestors of one group of bacteria, known as desulfovibrios. This group of bacteria can breathe sulfate, emitting noxious sulfur gases. They generate ATP, the energy molecule, during the conversion of sulfate to sulfide. These bacteria removed sulfate, releasing hydrogen sulfide, the same smelly gas that today gives salt marsh mud and some hot springs their "rotten egg" odor. As they breathed sulfate, the desulfovibrios synthesized a kind of molecule called a porphyrin ring which passed electrons and generated ATP along the way. The metabolic talents of the desulfovibrios were never lost. Their ability to make porphyrins has been retained, reused, and modified by all sorts of life forms, including human beings. Related porphyrin molecules, bright red in color, circulate in our blood today, where they carry oxygen to our cells.

With porphyrin ring production in their repertoire, many types of bacteria evolved the ability to use the most reliable and abundant source of energy around: sunlight. When any molecule absorbs light, its electrons are boosted to a higher energy state. Usually the energy is simply dissipated as light or heat until the molecule returns to its normal state. But when the molecules are bound to porphyrins attached to proteins embedded in membranes as electron-transport chains, light energy can be retained and put to use. Light energy in many sorts of bacteria can be retained for conversion to ATP energy. ATP energy is then used for movement and synthesis, such as conversion of carbon dioxide from the atmosphere into the food and replicating carbon compounds needed to self-maintain and grow. This process of getting food from light and air—photosynthesis—freed some kinds of bacteria completely from their dependence on preformed organic compounds.

The evolution of photosynthesis is undoubtedly the most important single metabolic innovation in the history of life on the planet. It occurred not in plants but in bacteria. Early photosynthesis was different from that found in plants today. The first photosynthetic organisms were bacteria that used hydrogen gas or hydrogen sulfide for the process and never produced oxygen. The sun-loving bacteria took the hydrogen needed to combine with carbon directly from the atmosphere. Hydrogen was prevalent in the Archean Eon, when the sun had just recently "turned on" and begun to waft huge quantities of it spaceward. Hydrogen sulfide, a gas produced by volcanoes, was also more abundant on the early Earth than it is now. Over time, as atmospheric hydrogen was depleted, even disappearing altogether from local niches, more photosynthetic bacteria used hydrogen sulfide produced as wastes by fermenting and sulfide-breathing microbes. Photosynthetic bacteria—then as now—used light energy to cleave off the hydrogen molecules. They excreted unused yellow pellets of congealed sulfur. Their modern-day descendants, known as the green sulfur and purple sulfur bacteria, still carry on in this way.

Needing light, the microbes that could move to maximize their exposure had an advantage. Behavior began. Even in these very ancient times, a combination of movement and simple systems of chemical sensing developed for detecting foods and avoiding poisons.

The origin of rapid motion in bacteria seems to be connected to a rotary device that is unknown in cells with nuclei. A flagellum, or whiplike strand, is attached to the disk-shaped base of the bacterium. The round base, known as a "proton motor," actually spins around, propelled by changes of electric charge. Since the flagellum, made of flagellin proteins, is attached to this organic wheel, it naturally goes round, too. Usually the wheel and flagellum are external to the prokaryote. But in some bacteria, such as the spirochetes, the flagellum is internalized. Under microscopes spirochetes appear as tiny animated corkscrews. The first spirochetes were a form of fermenting bacteria and probably evolved very early in the history of life. One of the most prevalent sorts of bacteria, oxygen-intolerant fast-moving spirochetes are particularly interesting and will reenter our story later on.

Those bacteria that evolved rapid movement, like those human beings with automobiles, had an advantage. More access to different locations meant more opportunities. Just as you do not have to learn a new trade if you can travel by car to a place offering your old job, so fast-moving microbes did not need to develop new metabolisms. They simply swam to locales that were rich in the nutrients they liked, and employed their same old metabolism. Fast movers also had more exposure to foreign genes and organisms and thus were apt to evolve intricate symbioses in a wide variety of environments.

Microbes using light could not bury themselves in the mud for protection from the still scorching ultraviolet radiation, which would remove them from the very light they required. Instead, they sought filters. They lived in solutions rich in certain salts, such as sodium nitrate, or they seeped under sand or other substances that absorb ultraviolet but let in visible light. They "tanned" themselves by developing pigments that absorbed the harmful rays. And they created shade by building colonies so vast they altered the landscape of the visible world.

From the beginning of time on the watery earth, colonies of different types of microbes formed cooperative dwellings. These become visible as tacky scums, purple and sienna patches, and strange, multilevel pastures. Generation after generation of bacteria in the topmost layers died from radiation exposure, but their remains shielded the lower layers, which accumulated sand and sediment to form a sort of living rug. Microbial mats and muds dominated the low-lying watery Archean landscapes. They can still be found today—fibrous patches so thick they can be sliced with a knife—along the warm seacoasts of Baja California, Mediterranean Spain, the Persian Gulf, western Australia, and even seasonally along the Atlantic coast from Nova Scotia to South Carolina. Unremarkable in appearance and overlooked by those seeking grander scenery, these mats are living examples, virtually untouched by time, of an ancient, purely bacterial empire.

Besides protective filtering, there was another excellent strategy for surviving the damaging rays of the ancient sun. The development of mechanisms for repairing sun-damaged DNA turned out to be a most powerful tool for the building of the microcosm as well as the rest of the biosphere that arose from it. For example, a common mode of destruction of bacterial DNA by ultraviolet light is the production of "thymine dimers." Thymine, instead of pairing with its complement, adenine, becomes chemically confused and pairs with itself, entangling the DNA molecule to the point of uselessness. Death will ensue unless repair enzymes fix up the mess. Such repair enzymes remove the disabled portion—the thymine dimer—and copy new, healthy DNA to replace it. In other words, bacteria threatened by ultraviolet radiation had just developed DNA splicing, the mechanism that is exploited today in the laboratory under the rubric of genetic engineering. Nearly all organisms today still have repair enzymes, even though life has been shielded from harmful ultraviolet rays by an atmospheric ozone layer for 2,000 million years. In many bacteria to this day, repair enzymes must be activated by light.

The pressure to patch up damaged DNA or die induced the development of DNA repair systems. Sometimes instead of using healthy copies of their own genetic material, crowded bacteria borrowed DNA from their neighbors. In modern bacteria, bits of genetic information in the form of various DNA fragments are passed among different strains of bacteria. Although exchange is easiest between bacteria that are metabolically similar, any strain can potentially receive genes from any other through a succession of intermediaries. This allows genetic information to be distributed in the microcosm with an ease and speed approaching that of modern telecommunications—if the complexity and biological value of the information being transferred is factored in. By trading genes, bacterial populations are kept primed for their role in their particular environment and individual bacteria pass on their genetic heritage.

By adapting to life under harsh light, the microcosm had invented sex. Though this first sex was different from the kind of sex animals are involved in, it was sex all the same. Sex, as recognized by biologists, is the mixing or union of genes from separate sources. It is not to be equated with reproduction, since an old organism can receive new genes and thereby have sex without reproducing itself. Sex always involves at least one live organism, but the second source of genes does not have to be alive; it can be a virus or even DNA in a test tube.

On the early Earth there came a time when a bacterium replaced some of its sun-damaged genes with fresh ones from a virus, a live bacterium, or even the old, discarded DNA of a dead cell. That bacterium had sex. More fluid and more frequent than the meiotic "sperm-and-egg" sex of animals, which is locked into the process of reproduction, bacterial sex immeasurably intensified the complexity of the microcosm. Because bacteria may mix genes at any time and are not confined to doing so during reproduction, they are far more genetically promiscuous than animals.

Since bacterial gene-transfer does not depend on reproduction, it takes a little explaining to get used to. At the beginning of the bacterial sex act there are two partners. At the end there usually is only one sexually produced offspring: the parent itself—the recombinant bacterium that now carries genes from two sources. The bacterium, without even reproducing, may now carry 90 percent new genes. More casual and more immediately required for survival in otherwise hostile environments, this first kind of sex is really quite different from the animal and plant sort of sex which is tied to reproduction. Bacterial sex preceded animal sex by at least 2,000 million years and, like a trump card, it permitted all sorts of microbes to stay in the evolutionary game.

CHAPTER 5

Sex and Worldwide Genetic Exchange

THE first sort of sex to appear on our planet, the bacteria-style genetic transfer exploited by genetic engineers, was and is very important to life's most basic functioning. Although some go so far as to argue that sex predated the origin of cells, it must at least have evolved by sometime in the Archean Eon, over 3,000 million years ago. Indeed, the prokaryotic brand of sex weaves through the story of life on Earth and provides a key to understanding its history. Bacteria-style sex would be important later as a way of genetically "locking" together symbiotic mergers between very different organisms. And it was always crucial to the biota's "reaction time": its ability to respond quickly to environmental changes and emergencies.

Broadly defined, sex is simply the recombination of genes from more than one source. Because, in our vertebrate world, such gene recombination is vital to the process of reproduction by way of male and female gametes (sperm and eggs), we have difficulty separating the concept of sex from our style of reproduction. Actually, sexuality is not required for reproduction in most members of four out of the five kingdoms of living things.

Bacteria reproduce asexually by growing to twice their size, replicating their single strand of DNA, and then dividing, with one copy of the DNA in each new offspring cell. Alternatively, they may bud: a small cell containing a complete set of genetic material forms on the parent, gradually growing to adult size when it breaks off. They may also encase the DNA in a spore which can survive long periods of dryness or other adverse conditions, germinating again when conditions become wetter or more generally favorable.[20]

In any case, exchange of genes goes on quite independently of these reproductive processes and is not in the least required for them. Indeed, the vertebrate, mammalian form of sex is a rarity in the living world. Since it is stuck to and dependent upon the process of reproduction, it is not nearly so important to the biosphere as bacteria-style sex. Bacteria-style sex, which does not have to wait for reproduction but can occur at virtually any time, is the common thread tying together many observations from fields such as bacteriology, medicine, and molecular biology—observations which at first sight might seem unrelated.

Bacteriologists have long observed that microorganisms freely pass hereditary traits from one to another. As far back as 1928, Frederick Griffith discovered that live pneumococci could acquire traits from another type of pneumococci, even though the latter organisms were dead. In 1944 Oswald T. Avery, Colin MacLeod, and Maclyn McCarty published a shocking discovery that was so important and so different from what all their colleagues really believed that it was quietly received and not appreciated until later.[21] These men proved beyond a doubt that it was DNA and not protein that was responsible for the hereditary acquisition of the traits from the dead pneumococcus DNA. That is to say, even though they were dead, the disease-causing cells had sex and transferred their genes as long as their DNA stayed intact. These New York scientists showed that the chemical behind this "transformation" was nucleic acid: deoxyribonucleic acid, DNA. Understanding what it was about this DNA from dead organisms that made it capable of transforming the inherited features of bacteria became the obsession of a young graduate student from Chicago named James D. Watson. Along with the brilliant Francis Crick, Watson solved the riddle of the DNA's chemical structure. This simultaneously revealed the secret of its function. Thirty years later, the structure and function of DNA has become the focus of an entire field of science, molecular biology, that boasts thousands of workers.

The astonishing discoveries of how the double helix is formed and what it can do, and of how organic molecules replicate and mutate, dazzled the scientific and popular imagination alike. In addition, medical conquests of various specific disease-causing bacteria occupied, as always, another bright limelight. Observations of other microbial behavior occurred outside the glare of publicity and in relative isolation, with few people seeing the discoveries of molecular biology, bacteriology, and medicine as part of the single phenomenon of life in the microcosm.

Each discipline had explored and sometimes exploited the fact that different strains of bacteria in nature are constantly exchanging bits and pieces of their genetic material among one another in a more or less random fashion. The number of permanent genes in each bacterial cell can be 300 times fewer than that of a typical nucleated cell (such as those of which we are made). But weakness can be strength, and the reverse side of this genetic shortcoming is that it makes the bacterial world incredibly more flexible than our world of nucleated cells in terms of adaptation. A bacterium possesses only a streamlined, bare-bones minimum of instructions for replication and maintenance. Any extra adaptive abilities that would enable the bacterial strain to survive in specialized conditions are conferred by visiting genetic particles known as small replicons. These small replicons, by means of bacterial sex, visit from cell to cell. Sometimes they become integrated into the main bacterial DNA, known as the genophore; at other times they exert an influence as a sort of genetic satellite, apart from the genophore.

All this is not to say that a single bacterium will necessarily benefit from a visiting replicon. But various sorts of replicons can represent significant percentages of the genetic makeup of a bacterium; their influence per cell stands a chance of being far greater. Genetic transfers differ in bacteria because they do not need to wait for reproduction. Bacteria are not altered by an automatic addition of 50 percent new genes, which is always what happens to sexually reproducing animals and plants when they produce offspring. Paradoxically from our standpoint in the human world, prokaryotes have sex more often with more partners yet they remain more faithful in terms of the degree of similarity shown between parents and offspring.

Human beings and most plants and animals simply cannot make dramatic changes in their outward appearance or metabolism by receiving visiting genes that code for at most only a few percent of their proteins. The huge numbers of synchronized interconnected cells in animals and plants compromise our genetic potential. In the fast world of bacterial genetic switching, large organisms are cumbersome, overstuffed operators. They can only trade some of their genes some of the time. They have great restrictions on the nature of their sexual partners. People and other eukaryotes are like solids frozen in a specific genetic mold, whereas the mobile, interchanging suite of bacterial genes is akin to a liquid or gas. If the genetic properties of the microcosm were applied to larger creatures, we would have a science-fiction world in which green plants could share genes for photosynthesis with nearby mushrooms, or where people could exude perfumes or grow ivory by picking up genes from a rose or a walrus.

Yet in the microcosm, the implications of easy genetic exchange are even more staggering. For if, indeed, all strains of bacteria can potentially share all bacterial genes, then strictly speaking there are no true species in the bacterial world. All bacteria are one organism, one entity capable of genetic engineering on a planetary or global scale. In the words of Canadian bacteriologists Sorin Sonea and Maurice Panisset, this entity is, in effect "a unique, complex type of clone, composed of highly differentiated (specialized) cells."[22]

The variety of mechanisms for these constant, opportunistic exchanges allows for an amazing number of options. In addition to its main strand of DNA, which bacteriologists call a genophore, chromoneme, or large replicon, a bacterial cell contains at any given time a number of self-replicating DNA molecules, the small replicons, visiting from the cells of other bacterial strains. Some of these small replicons can insert themselves into the cell's main DNA strand where they can divide at the same time as the main strand or by themselves at a different time. Other small replicons float around in the cell body or attach to its membrane, directing the synthesis of proteins for various purposes. If the trait conferred by the small replicon is useful, the gene for it may be retained by a jump over to the large replicon. Likewise, part of the large replicon may be eliminated when outmoded in any given environment.

In addition to small replicons, passive nonreplicating fragments of DNA can enter a cell if it has an appropriate receptor site and replace a similar fragment on the cell's large replicon, thus altering the genetic message. Some such DNA fragments, called transposons, are equipped with linking sequences that allow them to insert themselves freely into any replicon, large or small.

Small replicons are known in their various forms and disguises as plasmids, episomes, prophages, phages, and the devilish particles that plague our nucleated cells, viruses. Each type generally exhibits different characteristics or a different modus operandi, though there is some overlap. Most small replicons bring with them an assortment of "tools": genes that enable them to replicate and transfer themselves to other cells, or that enable the host cell to express the information they bring. For example, a prophage inside a bacterium can replicate many times, and to become a virus it can encase each of its copies in a protein shell that later is used to attach the copy to other bacteria. The host bacterium, full of mature viruses, then "lyses" or disintegrates, releasing the lytic phages, as they are now called, to infect other bacteria, sometimes after wind and water currents have borne them far from their origins.

In the process known as transduction, a fragment of the bacterium's DNA or any other small replicon present can be taken up into the protein case and travel to other bacteria. And in conjugation, moving DNA causes a tiny tube, or pilus, to form between two cells. The donor transfers a copy of its DNA through the tube to the recipient.

Transduction and conjugation are the bacterial world's chief methods of sharing immunity to drugs. The speed with which hereditary drug resistance spreads among bacterial communities impressively demonstrates the power and efficiency of their communications network. For example, the gene that directs the synthesis of an enzyme that digests penicillin probably originated in soil bacteria. Via exchanges by phages it eventually reached the staphylococci in our hospitals, many of which are now frighteningly resistant to penicillin. It has been estimated that a eukaryotic organism needing a similar enzyme might have had to wait about a million years for random mutations to produce it. On the other hand, the spirochete that causes syphilis is still susceptible to penicillin, possibly because it has not received from other bacteria genes that break penicillin down.

Its minimal number of genes leaving it deficient in metabolic abilities, a bacterium is necessarily a team player. A bacterium never functions as a single individual in nature. Instead, in any given ecological niche, teams of several kinds of bacteria live together, responding to and reforming the environment, aiding each other with complementary enzymes. The various kinds of bacteria in the team, each present in enormous numbers of copies, coordinate the release of their enzymes according to the stages in a task. Their life cycles interlock, the waste products of one kind becoming the food sources of the next. Intricately meshed in this way, bacteria occupy and drastically alter their environments. In huge and changing numbers, they perform tasks of which individually they are incapable.

With other strains of bacteria always living nearby, ready to contribute useful genes or metabolic products, and to reproduce under favorable conditions, the team's overall efficiency is kept in top condition. Over time this natural selection makes the team quite stable and capable of maintaining a complex group metabolism. Bacterial teams are to the Earth what our internal organs are to us. Dispersed, as are our blood cells, in any viable corner of our planet, the composition of a given team becomes adjusted for local conditions. The arrangements are dynamic, ready to change or to start in a new way if conditions around them change. Sometimes—for example, in slowly burning coal piles—the teams even maintain their own microclimate complete with temperature control. In addition, throughout the biosphere bacterial teams interact with various plants, animals, and fungi in the eukaryotic world. These larger consortia also operate with the dynamic harmony of a single organism.

The tasks undertaken by bacterial teams amount to no less than the conditioning of the entire planet. It is they that prevent all once-living matter from becoming dust. They turn us into food and energy for others. They keep the organic and inorganic elements of the biosphere cycling. Bacteria purify the Earth's water and make soil fertile. They perpetuate the chemical anomaly that is our atmosphere, constantly producing fresh supplies of reactive gases. The British atmospheric chemist James Lovelock suggests that certain microbially produced gases act as a control system to stabilize the living environment. Methane, for example, may act as an oxygen regulation mechanism and ventilator of the anaerobic (oxygenless) zone, whereas ammonia—another gas that reacts strongly with oxygen and therefore must be continually resupplied by microbes—possibly plays a major role in determining the alkalinity of lakes and oceans. A so-called greenhouse gas (like carbon dioxide) that lets in more radiation than it lets out, thereby increasing the temperature of the planet, ammonia also may have been important in the control of the ancient climate. Methyl chloride, a trace gas in the Earth's atmosphere, may regulate the ozone concentration of the upper atmosphere, which in turn has an effect on the amount of radiation which reaches the surface. This influences the further growth of gas-producing microbes. The list goes on and on. The environment is so interwoven with bacteria, and their influence is so pervasive, that there is no really convincing way to point your finger and say this is where life ends and this is where the inorganic realm of nonlife begins.

The bacterial world has, in effect, retained the time when the organism was not dependent on the careful packaging of DNA that occurs in nucleated organisms, but had many options open. Bacterial DNA and RNA did not become trapped in the nuclei of sexually reproducing species but remained flexible in a modular fashion. In the bacterial world, free-lancing DNA fragments, hovering between life and nonlife, constitute a powerful repertoire of tools for the ongoing business of evolution. When a virus enters our nucleated cells and issues its strange instructions, it can wreak havoc. For bacteria, however, and for the parts of our cells that are most like bacteria, viral and other DNA intrusions are routine and on the whole adaptive. New combinations are continually being tried out in the microcosm, and the evolution of these changes feeds the evolution of the nonbacterial layers of the biota as well.

How limited and rigid life becomes, in a fundamental sense, as it extends down the eukaryotic path. For the macrocosmic size, energy, and complex bodies we enjoy, we trade genetic flexibility. With genetic exchange possible only during reproduction, we are locked into our species, our bodies, and our generation. As it is sometimes expressed in technical terms, we trade genes "vertically"—through the generations—whereas prokaryotes trade them "horizontally"—directly to their neighbors in the same generation. The result is that while genetically fluid bacteria are functionally immortal, in eukaryotes, sex becomes linked with death.

With teams of growing and dying bacteria inhabiting every possible locale on the Earth's surface and continually selecting the best local solutions to the problem of maintaining life at any given time, the surface is kept in a stable and hospitable state. Consisting of life, the environment is continually regulated by life, for life. The supreme success of gene-pooling in ensuring the maximum number of bacteria results in heightening the quantity of all life forms, including animals and plants, and in an acceleration of all biological cycles. Moreover, in their alliance with animals and plants, which could not live or evolve without them, the Earth's bacteria form a complete planetary regulatory system,[23] one that has the specific effect of stabilizing the percentages of reactive atmospheric gases and the general result of keeping the Earth habitable. People may never truly appreciate this feat of genetic engineering until explorers try settling Mars or making satellites habitable on a continual basis—which is what bacteria have been doing to the Earth throughout its long history.

Sonea and Panisset liken the teeming activities of the planetary bacterial superorganism to the functions of a computer with an enormous data bank—the bacterial genes—and a global communications network that processes "more basic information than the brain of any mammal."[24] They point to the worldwide spread of resistance to antibiotics as spectacular proof that "bacteria act as a united entity capable of solving complex problems, and solve them efficiently every time." Having evolved with and been prompted by the bacterial superorganism, human intelligence uses many similar techniques to solve problems and transmit information. Sonea and Panisset offer two analogies: Like microorganisms, humans may have many tools and know how to use them, but do not carry them around all the time. And humans can transmit information to both their children and their neighbors, just as bacteria transmit genes both vertically to their offspring and horizontally to other bacteria. As a result, humanity and the microcosm both maintain a continually growing pool of technical know-how. Acquired knowledge does not die with each individual or generation.

Our first awareness of the microcosm was hostile. Plagues, leprosy, and venereal diseases during the Middle Ages forced thinkers to recognize, at least intuitively, the phenomenon of worldwide genetic transfer. The bacterial lifestyle was described when and where it threatened human society. The old ideas of being "unclean" or even possessed, and the post-medieval discoveries of cleanliness and antibiotics in fighting illness represent the front on which the new human intellect and the old planetary microcosm first met. Louis Pasteur (1822–1895), who proved the microbial origin of such devastating diseases as foot and mouth disease, plague, and wine rot, set the tone of the relationship from the start. The context of the encounter between intellect and bacteria defined medicine as a battleground: bacteria were seen as "germs" to be destroyed. Only today have we begun to appreciate the fact that bacteria are normal and necessary for the human body and that health is not so much a matter of destroying microorganisms as it is of restoring appropriate microbial communities. Only today have we begun to appreciate the benign side of infection, the inheritance of desired characteristics which is exploited in genetic engineering.

For sheer scope, human information systems have only just begun to approach the ancient bacterial systems which have been trading bits of information like a computer network with a memory accumulated over billions of years of continuous operation. As we move from a purely medical view of microbes to an understanding of them as our ancestors, as planetary elders, our emotions also change, from fear and loathing to respect and awe. Bacteria invented fermentation, the wheel in the form of the proton rotary motor, sulfur breathing, photosynthesis, and nitrogen fixation, long before our evolution. They are not only highly social beings, but behave as a sort of worldwide decentralized democracy. Cells basically remain separate, but can connect and trade genes with organisms of even exceedingly different backgrounds. Realizing that human individuals also remain basically separate but can connect and trade knowledge with very different others may be taking a step toward the ancient wisdom of the microcosm.

Although we are just learning of its existence, the microcosm has much more in store for us. The move from an industrial society of coal and steam engines to a postindustrial society of televisions and computers has been compared to the difference between brains and brawn. So it is with the resources of the Earth. As we have exploited the trapped energy of fossil fuels such as coal, oil, and natural gas millions of years old, we may be able to tap into information resources *billions* of years old. The molecular microelectronics of photosynthesis, genetic engineering, embryo development, and other natural technologies lie waiting for us. Gaining access to such stored information, becoming fluent in such mysteries, will lead to changes so staggering they are beyond what we can now know.

On the eve of the Proterozoic Eon 2,500 million years ago, every likely nook of the Earth's surface was teeming with bacteria. Life had responded to ultraviolet radiation with sex, and this benign infectiousness, this germ of intelligence, had helped life spread its useful inventions. By the Proterozoic Eon thousands of metabolic devices had been invented, indeed all the major ones known today. For example, bright orange or purple carotenoid pigments and vitamin A, compounds protecting against the treacherous effects of bright light, were in the possession of certain bacteria. In the modern world carotenoids color carrots orange, and vitamin A is now a chemical preliminary in the body's process of making rhodopsin, the visual pigment of our eyes. Many such inventions of the microcosm have never been lost.

To a casual observer, the early Proterozoic world would have looked largely flat and damp, an alien yet familiar landscape, with volcanos smoking in the background and shallow, brilliantly colored pools abounding and mysterious greenish and brownish patches of scum floating on the waters, stuck to the banks of rivers, tinting the damp soils like fine molds. A ruddy sheen would coat the stench-filled waters. Shrunk to microscopic perspective, a fantastic landscape of bobbing purple, aquamarine, red, and yellow spheres would come into view. Inside the violet spheres of *Thiocapsa*, suspended yellow globules of sulfur would emit bubbles of skunky gas. Colonies of ensheathed viscous organisms would stretch to the horizon. One end stuck to rocks, the other ends of some bacteria would insinuate themselves inside tiny cracks and begin to penetrate the rock itself. Long skinny filaments would leave the pack of their brethren, gliding by slowly, searching for a better place in the sun. Squiggling bacterial whips shaped like corkscrews or fusili pasta would dart by. Multicellular filaments and tacky, textilelike crowds of bacterial cells would wave with the currents, coating pebbles with brilliant shades of red, pink, yellow, and green. Showers of spores, blown by breezes, would splash and crash against the vast frontier of low-lying muds and waters.

All beings in this bygone biosphere were prokaryotic—that is, they all lacked nuclei. Their genes were not packed into chromosomes wrapped by a nuclear membrane. They had evolved and distributed probably all the major metabolic and enzymatic systems. Their cycling of gases and soluble compounds through the Earth's atmosphere and water had created the fundamentals of the planetary ecosystem. Although the oxygen revolution to come was to drive these Archean anaerobes underground and under water, many bacteria living during this time have survived essentially unchanged for more than three billion years.

CHAPTER 6

The Oxygen Holocaust

THE oxygen holocaust was a worldwide pollution crisis that occurred about 2,000 million years ago. Before this time there was almost no oxygen in the Earth's atmosphere. The Earth's original biosphere was as different from ours as that of an alien planet. But purple and green photosynthetic microbes, frantic for hydrogen, discovered the ultimate resource, water, and its use led to the ultimate toxic waste, oxygen. Our precious oxygen was originally a gaseous poison dumped into the atmosphere. The appearance of oxygen-using photosynthesis and the resulting oxygen-rich environment tested the ingenuity of microbes, especially those producing oxygen and those nonmobile microorganisms unable to escape the newly abundant and reactive gas by means of motion. The microbes that stayed around responded by inventing various intracellular devices and scavengers to detoxify—and eventually exploit—the dangerous pollutant.

The unceasing demand for hydrogen initiated the crisis. Life's need for carbon-hydrogen compounds had already almost depleted carbon dioxide from the atmosphere. (The atmospheres of Mars and Venus today are still more than 95 percent carbon dioxide; the Earth's is only 0.03 percent.) The lighter hydrogen gas kept escaping into space where it reacted with other elements, becoming ever less available. Even the Earth's hydrogen sulfide, gurgling up through volcanoes, was becoming insufficient to supply the vast communities of photosynthetic bacteria that by the late Archean dominated the soils and waters.

But the Earth was still full of an abundant hydrogen source: dihydrogen oxide, a.k.a. water. Until now, the strong bonds between the hydrogen and oxygen atoms in the water molecule (H_2O)—much stronger than those holding together the two hydrogens in hydrogen gas (H_2), hydrogen sulfide (H_2S), or organic molecules (CH_2O)—had been unbreakable by the resourceful, hydrogen-craving bacteria. Sometime after photosynthesis in the oxygen-poor atmosphere of the early Earth had been well established, however, a kind of blue-green bacteria solved the hydrogen crisis forever. These were the ancestors to modern cyanobacteria.

The cyanobacterial ancestors seem to have been mutant sulfur bacteria desperate to continue living as their store of hydrogen sulfide dwindled. These organisms were already photosynthetic, and already had proteins inside them organized into so-called electron transport chains. In some of the blue-green bacteria, mutant DNA which coded for the electron transport chains duplicated. Experts at capturing sunlight in their reaction center to generate ATP, the new DNA led to the construction of a second photosynthetic reaction center. This second reaction center, by using light-generated electron energy from the first center, absorbed light again; but this was higher-energy light, absorbed at shorter wavelengths, that could split the water molecule into its hydrogen and oxygen constituents. The hydrogen was quickly grabbed and added onto carbon dioxide from the air to make organic food chemicals, such as sugars. In an evolutionary innovation unprecedented, as far as we know, in the universe, the blue-green alchemists, using light as energy, had extracted hydrogen from one of the planet's richest resources, water itself. This single metabolic change in tiny bacteria had major implications for the future history of all life on Earth.

The new dual light-powered system not only generated more ATP but accessed an almost inexhaustible hydrogen source; with it, the first cyanobacteria were spectacularly successful. Colonizing every available spot that guaranteed sunlight, carbon dioxide, and water, they spread over the Earth's surface. Today they grow like weeds on rocks, swimming pool surfaces, drinking fountains, shower curtains, and sand flats—wherever there is water and light. When the Lascaux caves of southern France, whose walls bear paintings made by paleolithic hunters, were opened to the public in the late 1970s they were soon forced to close again. Once light and water were let in, the cyanobacteria grew and divided, threatening the dazzling 40,000 year-old cave paintings by covering their surfaces.

On the early Earth the blue-green tint crept over the minerals and muds. It glided, grew, crept, and swelled, gradually expanding along river shores and meteoritic rubble, upon volcanic debris and in puddles. Like all rapidly growing living systems, the cyanobacteria produced prodigious amounts of waste. Whereas their ancestors had taken in hydrogen sulfide (H_2S) and released sulfur (S), they took in water (H_2O), and released oxygen gas(O_2). The toxicity of uncombined oxygen is well established and the new oxygen gas produced by the photosynthetic colonies immediately threatened them most, since they were closest to the source. Oxygen—a deadly poison to all early life—bubbled up from the mats and muds, polluting the marshes, pools, and riverbeds.

Oxygen is toxic because it reacts with organic matter. It grabs electrons and produces so-called free radicals: highly reactive, short-lived chemicals that wreak havoc with the carbon, hydrogen, sulfur, and nitrogen compounds at the basis of life. Oxygen breaks down or renders useless the small metabolites—food—that otherwise become components in cellular systems. Oxygen combines with the enzymes, proteins, nucleic acids, vitamins, and lipids that are vital to cell reproduction. Oxygen also quickly reacts with atmospheric gases, including hydrogen, ammonia, carbon monoxide, and hydrogen sulfide. In a word, oxygen burns: it "oxidizes," dramatically changing soil minerals such as iron, sulfur, uranium, and manganese to oxidized forms such as hematite, pyrite, uraninite, and manganese dioxide—that is, to new, oxygen-bound compounds of these metals.

The biosphere at first could absorb the oxygen pollution. As long as there were abundant metals and gases that could react with it, oxygen did not build up in the atmosphere. Moreover, the production of oxygen probably varied with the season—more in summer when photosynthetic activity was the greatest, less in winter. Some photosynthetic microbes must have been able to alternate between oxygen-producing and oxygenless photosynthesis, depending on whether hydrogen and hydrogen sulfide were in good supply. (As in the beautiful *Oscillatoria limnetica*, oxygen production was at first optional. Discovered by Professor Yehuda Cohen of Eilat Marine Station in the hot water of Solar Lake in the Negev Desert in 1975, *O. limnetica* has a "chameleon" physiology. It can use hydrogen sulfide for photosynthesis when that compound is plentiful, in which case the cyanobacterium gives off no oxygen. But when starved for hydrogen it switches to using the hydrogens from water, and the leftover oxygen is released as waste into the air.) The amount of poisonous oxygen fluctuated with season, volcanic activity, cyanobacterial population, and many other variables.

As paleobiologist J. William Schopf has pointed out, "The paleobiological record shows . . . that the advent of oxygenic photosynthesis was the singular event that led eventually to our modern environment."[25] Although the mineral record clearly shows a sudden build-up in the amount of atmospheric oxygen, exactly when oxygen-producing photosynthesis left earth-changing quantities of oxygen in the atmosphere is hotly debated.

An intriguing sign that oxygen production existed long before most think it did is displayed in the same ancient Isua rocks that held concentrations of graphite signaling the possible remains of photosynthetic bacteria. Some beautiful banded rock formations made of alternating stripes of different types of iron oxides—oxidized hematite and less oxidized magnetite—exist there. In some places the alternating bands are only microns, rather than meters, in size. These banded iron formations (BIFs) are important to us because they supply our raw sources of mineable iron. In fact fewer than twenty BIFs, all dating from the Proterozoic Eon, make up more than 90 percent of the world's commercial iron supply. For such BIFs to occur, both large bodies of water and fluctuating amounts of oxygen are thought to have been needed. Oxygen-producing photosynthetic bacteria may have thrived on the surface of warm, volcanic pools along vents and rifts in iron-rich waters, their seasonal bursts of growth and accompanying puffs of oxygen waste producing the colorful mineral layers.

Photosynthetic bacteria may also have had collaborators in the making of banded iron. Certain iron-oxidizing or pipe-clogging bacteria can take oxygen from the environment and derive energy in nutrient-poor waters by combining it with iron. The combination of oxygen and iron leads to a chemical reaction which makes rust. These bacteria extract energy from the chemistry of that reaction and the rust gets deposited on their long and fibrous bodies. In an oxygenless world such bacteria probably proliferated both above and below the zones of oxygen production. Year after year they could have scavenged oxygen at the edges of the cyanobacterial communities, precipitating it as rust. Iron-oxidizing bacteria may have helped form the vast quantities of ancient iron ore. And the alternating bands of iron ore may be a record of their ancient relationship with cyanobacteria: hematite would have been produced in summer, when the cyanobacteria produced more oxygen, leading to rustier iron, whereas the magnetite layers probably built up in the winter, when photosynthetic oxygen production and therefore iron oxidation were at a lull.

During a part of the Proterozoic Eon from 2,200 to 1,800 million years ago, there was a great burst of banded iron formation unparalleled since. But more than 3,000 million years ago? If microbial life is indeed implicated in the making of banded iron, these glittering bands of metal ore in Labrador and Greenland represent the naissance of bacterial communities and would be the oldest evidence of bacterial oxygen-producing photosynthesis and, indeed, of life itself.

Another spectacular clue to the earliest aerobic activities of life is the accessibility of gold, the metal so prized throughout human history, that was brought up from the molten center of the earth sometime in the Archean Eon. Sediment-embedded gold is limited to several suites of rocks from Archean times, and what little there is worldwide is startlingly concentrated. The Witwatersrand mines in the Transvaal of South Africa, for example, account for some 70 percent of all the gold that has circulated throughout the history of civilization. Lesser deposits are located in northwestern Australia,

Elliot Lake in northern Ontario, and southern Russia—none approaching the yield of the Transvaal.

Gold miners descending in elevator shafts thousands of feet into the Earth are effectively going back to earlier times, past layers of volcanic ash and the debris of ancient rivers to older surfaces. To seek new gold deposits, they follow the carbon leader—a distinct layer of conglomeritic rock that contains a great deal of organic carbon in it. The carbon leader, trapped between limestones and shales, contains thin seams of pyrite, gold, and often even uranium ore. The Witwatersrand carbon leader also contains microscopic filamentous and ball-like structures that are inexplicable by mineralogy alone.

D. K. Hallbauer, a South African economic geologist, first related these structures to life by interpreting them as the fossilized components of lichens. Since lichens are a complex alliance of algae and fungi of which no fossil record appears until 2,000 million years after the South African gold was deposited, no one accepted Hallbauer's specks as lichens. A more convincing possibility is that filamentous and coccoid bacteria trapped detrital flakes of gold.

Flowing from the interior of a tectonically active Archean Earth, hot magma brought tiny amounts of the heavy, molten gold from the Earth's mantle to the surface in a finely dispersed form in rocks of iron, magnesium, and silicate. Because gold moves in and out of rocks more fluidly in the absence than in the presence of oxygen, it would have been readily eroded from the rocks by rivers and streams and carried to the sea. If it encounters high concentrations of oxygen and organic carbon, however, gold comes out of solution—it "flocculates." Any colonies of photosynthesizing bacteria dwelling along the river shores could have played a role in this. By producing quantities of oxygen and carbon-rich compounds, they may have coaxed gold from the water to come out in gooey flocs and be deposited along the banks and beds of the ancient waterways.

Certain bacteria (*Chromobacterium violaceum*) today produce cyanate, a chemical used by gold mining companies in the extraction of gold from carbon-rich sediment. Perhaps the ancestors of these microbes lived in the mineral-laden Archean rivers, consolidating dissolved gold into discrete particles. The oxygen and carbon from cyanobacteria itself may have been enough to precipitate the gold from solution. In southern Africa gold from the interior of the earth was laid along a river system estimated to be five times the size of the Mississippi. Eventually the great Witwatersrand river ways, draining huge amounts of water into the ocean 2,500 million years ago, dried out. The rivers were buried under kilometers of sediments and folded. They were not found until the last century, when Dutch-derived Afrikaner settlers of South Africa were attracted by the gold speckles of some dark rocks that had outcropped in the Transvaal desert. They followed the outcrop deep into the ground, finding the gold by tracing the buried ancient river system of which the carbon leader was a part.

The clearest evidence of the lives of ancient and extensive bacterial confederacies, however, are stromatolites. Stromatolites were to the Proterozoic landscape what coral reefs are to the present ocean: rich and beautiful collectives of intermingled, interdependent organisms. These domed, conical, columnar, or cauliflower-shaped rocks, found throughout the fossil record and still in existence today, are composed of rock layers that were once microbial mats. Communities of bacteria, especially photosynthetic cyanobacteria, lived and died atop one another. When some stromatolites, such as those in the hot springs in Saratoga, New York, were first described in the late nineteenth century by geologist Charles

Walcott and others, they were termed *Cryptozoa*, Greek for "hidden animals." Some of the ancient stromatolites exceeded thirty feet in height.

Today—in very restricted parts of the world—we can see that the top layers, only a few centimeters in width, are dominated by photosynthetic blue-green bacteria. This is the living part of the mat. Everything below is composed of dormant bacteria, chalk, sand, gypsum, and other debris bound together by the matrices of earlier mats. The top layer is horizontally striped. Beneath the top layer of photosynthesizers are thriving populations of anaerobic purple photosynthesizers, which are sulfur depositers. Beneath them are dependent microbes, living on the produce or the bodily remains of the others. Living stromatolites may be found today in the Persian Gulf, western Australia, and the Bahama Islands. Breaking them open reveals with a hand lens the gelatinous growth of many different kinds of bacteria, but the actual carbonate precipitation is still done by cyanobacteria. In soft bacterial mats and hardened ones (which are by definition stromatolites) bacteria grow in multicellular layers that are as complex and differentiated, in their own way, as the tissues of animals.

The oldest known stromatolites are about 3,500 million years old, their carbon-rich layers convincing evidence that photosynthetic microbial communities—aerobic or not—were thriving by that time. Though they are scarce in Archean rocks, they skyrocketed to success in the Proterozoic Eon, dominating the landscape.

For tens of millions of years excess oxygen was absorbed by live organisms, metal compounds, reduced atmospheric gases, and minerals in rocks. It began to accumulate in the atmosphere only by fits and starts. Many local populations were killed off, and many adaptations and protective devices evolved. From blue-green cyanobacteria that produced oxygen part-time emerged grass-green bacteria that emitted it continually. Thousands of species of aerobic photosynthesizers arose adapted to rocks, hot water locales, and scums. But by about 2,000 million years ago, the available passive reactants in the world had been used up and oxygen accumulated rapidly in the air, precipitating a catastrophe of global magnitude. People are gravely worried today about an increase in atmospheric carbon dioxide from 0.032 to 0.033 percent caused by our massive burning of fossil fuels. It is supposed that the "greenhouse effect" of the additional heat trapped by extra CO_2 could melt the ice caps, raising the level of the seas and flooding our urban coastlines, resulting in mass death and destruction. But the industrial pollution of our present, Phanerozoic Eon is nothing compared to the strictly natural pollution of Archean and Proterozoic times. About 2,000 million years ago—give or take a couple of hundred million years—oxygen started rapidly increasing in our atmosphere. The Archeo-Proterozoic world saw an absolutely amazing increase in atmospheric oxygen from one part in a million to one part in five, from 0.0001 to 21 percent. This was by far the greatest pollution crisis the earth has ever endured.

Many kinds of microbes were immediately wiped out. Oxygen and light together are lethal—far more dangerous than either by itself. They are still instant killers of those anaerobes that survive in the airless nooks of the present world. When exposed to oxygen and light, the tissues of these unadapted organisms are instantly destroyed by subtle explosions. Microbial life had no defense against this cataclysm except the standard way of DNA replication and duplication, gene transfer, and mutation. From multiple deaths and an enhanced bacterial sexuality that is characteristic of bacteria exposed to toxins came a reorganization of the superorganism we call the microcosm.

The newly resistant bacteria multiplied, and quickly replaced those sensitive to oxygen on the Earth's surface as other bacteria survived beneath them in the anaerobic layers of mud and soil. From a holocaust that rivals the nuclear one we fear today came one of the most spectacular and important revolutions in the history of life.

From the earliest days of local oxygen exposure, gene duplication and transfer resulted in many protective mechanisms. The new genes were as important as survival manuals. The information they contained, so valuable to life on the newly oxygenic Earth, spread through the reorganizing microcosm. Bioluminescence and the synthesis of vitamin E are some of the innovations scientists surmise arose in response to the oxygen threat. But adaptation didn't stop there. In one of the greatest coups of all time, the cyanobacteria invented a metabolic system that *required* the very substance that had been a deadly poison.

Aerobic respiration, the breathing of oxygen, is an ingeniously efficient way of channeling and exploiting the reactivity of oxygen. It is essentially controlled combustion that breaks down organic molecules and yields carbon dioxide, water, and a great deal of energy into the bargain. Whereas fermentation typically produces two molecules of ATP from every sugar molecule broken down, the respiration of the same sugar molecule utilizing oxygen can produce as many as thirty-six. Pushed—not, probably, to its breaking point, but to a point of intense global stress—the microcosm did more than adapt: it evolved an oxygen-using dynamo that changed life and its terrestrial dwelling place forever.

Some cyanobacteria respire only in the dark—apparently because they use some of the same molecular machinery for both their respiratory and their photosynthetic electron transport chains. The shared parts can't be used simultaneously for both pathways. (Algae and plants both respire and photosynthesize because the two processes take place in different parts of the cell: photosynthesis in the chloroplasts and respiration in the mitochondria. These two organelles are tantalizing hints to the evolutionary fate of two kinds of microbes—hints to be elaborated on in the next chapter.)

Cyanobacteria now had both photosynthesis which generated oxygen and respiration which consumed it. They had found their place in the sun. Given only sunlight, a few salts always present in natural waters, and atmospheric carbon dioxide, they could make everything they needed: nucleic acids, proteins, vitamins, and the machinery for making them. If biosynthetic ability alone were considered a measure of evolutionary advancement, we humans would be far behind the cyanobacteria. Our complicated nutritional requirements leave us utterly dependent on plants and microbes to supply what we cannot make for ourselves. We are, in a very real sense, parasites of the microcosm.

Not surprisingly, with the greater quantities of energy available to them, cyanobacteria exploded into hundreds of different forms, tiny (most only a few micrometers in diameter) and large (80 micrometers or eight percent of a millimeter). They made simple spheres embedded in a gelatinous matrix of sheets of many cells, elaborately branched filaments that could release wet spores from their tips, and cells containing special oxygen-proof cysts that carried on anaerobic nitrogen fixation.

They spread into greater extremes of the environment, from cold marine waters to hot freshwater springs. New food relationships developed as other bacteria fed off cyanobacterial starch, sugar, small metabolites, and even the fixed carbon and nitrogen of their dead bodies. But most significant, cyanobacteria's continuing air pollution forced other organisms to acquire the ability to use oxygen, too. This set off waves of speciation and the creation of elaborate forms and life cycles among them.

The stabilization of atmospheric oxygen at about 21 percent seems to be a mute consensus reached by the biota millions of years ago; indeed it is a contract still respected today. If the oxygen concentration had ever risen much higher than this, the fossil record certainly would reveal evidence of worldwide conflagration. The present high, but not too high, level of oxygen in our atmosphere gives the impression of a conscious decision to maintain balance between danger and opportunity, between risk and benefit. Even the rain forests and grasslands are extremely flammable when water levels are low. If oxygen were a few percent higher, living organisms themselves would spontaneously combust. As oxygen falls a few percent aerobic organisms start to asphyxiate. The biosphere has maintained this happy medium for hundreds of millions of years at least. While just how this works is still a mystery, we will see in our last chapter how worldwide regulatory mechanisms controlling temperature and gas composition can hypothetically arise from the normal growth properties of organisms. The leveling off and subsequent continuous modulation of the quantities of oxygen in the atmosphere was an event as welcome as the holocaust was terrible. One way of putting it is that, assuming life halted the oxygen build-up, it must have developed tremendous knowledge of antipollution engineering systems. The alternative perspective is that the cybernetic control of the Earth's surface by unintelligent organisms calls into question the supposed uniqueness of human consciousness. Microbes apparently did not plan to bring under control a pollution crisis of amazingly daunting proportions. Yet they did what no governmental agency or bureaucracy on Earth today could ever do.

Growing, mutating, and trading genes, some bacteria producing oxygen and others removing it, they maintained the oxygen balance of an entire planet.

Although it makes up only one-fifth of our atmosphere, from a chemical point of view atmospheric oxygen is extremely abundant: it should react with other chemicals to form stable compounds such as carbon dioxide and nitrate salts. As Jim Lovelock puts it, "The present level of oxygen tension is to the contemporary biosphere what the high-voltage electricity supply is to our twentieth-century way of life. Things can go on without it, but the potentialities are substantially reduced. The comparison is a close one, since it is a convenience of chemistry to express the oxidizing power of an environment in terms of its reduction-oxidation (redox) potential, measured electrically and expressed in volts."[26]

As soon as there were significant quantities of oxygen in the air an ozone shield built up. It formed in the stratosphere, floating on top of the rest of the air. This layer of three-atom oxygen molecules put a final stop to the abiotic synthesis of organic compounds by screening out the high-energy ultraviolet rays.

The production of food and oxygen from light were to make microbes the basis of a global food cycle that extends to us today; animals could never have evolved without the food of photosynthesis and the oxygen in the air. The energy dynamo created by cyanobacterial pollution was a prerequisite for a new unit of life—the nucleated cell which is the fundamental component of plant, animal, protist, and fungal life. In eukaryotes, genes are packaged in a nucleus and there is an elaborate orchestration of internal cell processes, including the presence in the area surrounding the nucleus (the cytoplasm) of mitochondria—special structures that metabolize oxygen for the rest of the cell. So different is the organization of the eukaryotic from the prokaryotic or bacterial cell that the two types represent the most fundamental separation among known life forms. Perhaps their origin in the midst of the extreme selection pressures of the oxygen catastrophe made eukaryotic cells so different. But the difference between nonnucleated bacterial cells and cells with nuclei is far greater than that between plants and animals.

Before cyanobacteria split water molecules and produced oxygen, there was no indication that the Earth's patina of life would ever be more than an inconspicuous scum lying on the ground. That it did develop and expand into gardens and jungles and cities is testimony to the power of microbial mats and seaside slime to alter each other in their local habitats. But the microbes created an even greater impact. They altered the entire surface of the Earth. The biosphere, humming with the thrill and danger of free oxygen, eventually emerged from the crisis. But the Earth was a changed place. It had become a planetary anomaly.

By the middle of the Proterozoic Eon 1,500 million years ago most of biochemical evolution had been accomplished. The Earth's modern surface and atmosphere were largely established. Microbial life permeated the air, soil, and water, cycling gases and other elements through the earth's fluids as they do today. With the exception of a few exotic compounds, such as the essential oils and hallucinogens of flowering plants and the exquisitely effective snake venoms, prokaryotic microbes can assemble and disassemble all the molecules of modern life.

Judging from the perspective of the planetwide accomplishments of early life, it is not surprising that the development of life's biochemical repertoire occurred over a full two billion years. The microbial stage lasted nearly twice as long as the rest of evolution to the present day. As Abraham Lincoln is reported to have said, "If I had eight hours to chop down a tree, I'd spend six sharpening my ax." The microcosm did just that. It set the stage for the respective evolutions of fungi, animals, and plants, all of which arose in relatively rapid succession. Just as in psychology, where the early years of infancy and childhood are known to be crucial to the development of adult personality, the early eons of life defined the contours of modern living. The Age of Bacteria transformed the earth from a cratered moonlike terrain of volcanic glassy rocks into the fertile planet in which we make our home. The alien primeval world lacking an atmosphere of oxygen was to be no more. Unlike neighboring Mars and Venus, whose atmospheres settled down to become stable chemical mixtures of carbon dioxide, the Earth had gotten energized. Delivered from the mercy of time, it became engulfed in the creative, autopoietic processes of life.

CHAPTER 7

New Cells

WITH the invention of aerobic or oxygen-using respiration, prokaryotes had tapped into an energy source far beyond their ability to fully exploit. Unaware of the global power they were generating, the respiring bacteria flourished in their local niches all over the globe for hundreds of millions of years. But as the level of atmospheric oxygen was rising up to 21 percent, perhaps about 2,200 million years ago when it still formed only a few percent of the atmosphere, a new kind of cell formed. This was the eukaryotic cell with its key feature, the nucleus, and its important secondary characteristic, oxygen-using cell parts known as mitochondria. When eukaryotes live as single cells they are called protists. Their fossils are known as acritarchs. As Stonehill College astronomer Chet Raymo points out, the difference between the new cells and the old prokaryotes in the fossil record looks as drastic as if the Wright Brothers' Kitty Hawk flying machine had been followed a week later by the Concorde jet.[27]

The biological transition between bacteria and nucleated

cells, that is between prokaryotes and eukaryotes, is so sudden it cannot effectively be explained by gradual changes over time. The division between bacteria and the new cells is, in fact, the most dramatic in all biology. Since plants, animals, fungi, and protists all are based on the nuclear design, the distinction reflects the common heritage of these organisms, which together form the superkingdom of the eukaryotes, which radically differs from the bacterial world, the superkingdom of the prokaryotes (kingdom *Monera*).

Not merely more advanced bacteria, the new cells were much larger and more complex. They bore circuitous channels of internal membranes, including the one enveloping the nucleus. Self-reproducing, neatly packaged organelles that used oxygen floated in their cytoplasm. The first new cells appear in the rock record as the first acritarchs, some 1,600–1,400 million years ago. These nearly featureless spherical microfossils with their thick walls are thought to be the resistant cysts of some kind of early algae. The more recent acritarchs, less than 1,000 million years old, are even larger and have sculptured, rather fancy outer coverings. Such acritarchs have now been taken from rocks in Scandinavia and from the early Proterozoic record of the Grand Canyon in Arizona, as well as from many other places.

If the acritarchs are really the remains of the earliest eukaryotes, they had nuclei separated from the rest of the cell by a membrane. Indeed, large cells show up clearly in certain light microphotographs taken of slices of Proterozoic rocks in all parts of the world. Inside their nucleated cells, the early eukaryotes probably had chromosomes. The DNA of chromosomes is intimately packaged with proteins, usually in a ratio of 40 percent DNA to 60 percent protein. In addition to all this protein, nucleated cells have as much as 1,000 times the amount of DNA found in bacterial cells. The function of such enormous quantities of DNA is one of the most intriguing puzzles of molecular biology. While some of the DNA is of course useful, much of it is so-called "redundant DNA," that is, copies of genes repeated elsewhere in the chromosomes.

Some of the unicellular eukaryotes or protists, such as the algae that left acritarchal cysts, also had chlorophyll-bearing packets capable of photosynthesis suspended in their cytoplasm. These photosynthetic cell parts, called plastids, coexisted in the algae or plankton cells with the oxygen-using parts, the mitochondria. Like the mitochondria, the plastids were capable of self-reproduction by direct division into two, despite being part of a larger protist cell. It is very possible, as we shall soon see, that plastids and mitochondria represent bacteria that got trapped inside other bacteria. The sudden and widespread appearance of acritarchs in the fossil record attests to the wild success of the new cells, probably intertwined communities of cells-within-cells, starting about 1,400 million years ago. Taking the form of the world's first marine plankton, the new cells floated and reproduced atop the ancient seas. Some of them died and were buried in the strange spherical and polygonal forms of fossil acritarchs.

The new cells seem to have been bacterial confederacies. They cooperated and centralized, and in doing so formed a new kind of cellular government. The upstarts were increasingly centrally organized, and their various cell organelles became integrated into a new biological unit. For example, in modern eukaryotes, the cytoplasm streams about inside the cell as though with a purpose. Such directed intracellular movement is never seen in bacteria. But most distinctively, instead of the bacterium's single replicon, eukaryotes have beaded, protein-bound DNA structures, the chromosomes, with their massive amounts of DNA. Many ideas have been put forward to explain the huge quantities of DNA in nucleated cells. The molecular biologists W. H. F. Doolittle and

Carmen Sapienza claim it is "selfish" DNA, there because replication is its lifestyle and it has been able to replicate and persist in the accommodating environment of the eukaryotic cell interior.[28] If this is the case, the redundant DNA need not have any biological function. Other scientists believe that the extra DNA provides a "reserve" of genetic information to be drawn upon by future generations, like money in the bank. Still others have proposed a kind of cellular predestination: future evolution, they say, is already encoded in the DNA, which will become useful with the passage of time.

We think it is highly unlikely that evolution is forward-thinking in this way. We suspect the repetitive DNA originally came from different bacteria—anaerobes, oxygen-users, and others—that came together in the community that became the eukaryotic cell. The extra DNA certainly was generated by the tendency of DNA to replicate, but it started from the DNA of the original replicating members of the merger. All kinds of extra copies of DNA were preserved not because of "selfishness," but because they were put to use in solving the problems of packaging and functioning those miniaturized wonders, the chromosomes. After all, the DNA of the nuclei of all the cells in your body, if lined up end-to-end, rather than folded and packed tightly as chromosomes, would stretch beyond the Earth to the moon and back, not once but 120,000 times.[29]

All cells either have a nucleus or do not. No intermediates exist. The abruptness of their appearance in the fossil record, the total discontinuity between living forms with and without nuclei, and the puzzling complexity of internal self-reproducing organelles suggest that the new cells were begotten by a process fundamentally different from simple mutation or bacterial genetic transfer. The scientific work of the past decade has convinced us that this process was symbiosis. Independent prokaryotes entered others. Inside them they digested cellular wastes; their waste, in turn, was used as food. The outcomes of such intimate sharing were permanent relationships, cells reproducing offspring well adapted to life within other cells. With time these populations of coevolved bacteria became communities of microbes so deeply interdependent they were, for all practical purposes, single stable organisms—protists. Life had moved another step, beyond the networking of free genetic transfer to the synergy of symbiosis. Separate organisms blended together, creating new wholes that were greater than the sum of their parts.

This theory of the origin of nucleated cells by symbiosis is by no means brand new. Like the Swiss geologist Alfred Wegener who first proposed the notion of continental drift simply by looking at the puzzle-piece shapes of the continents, biologists in the 1880s, using the newly invented compound light microscope, saw bacterialike inclusions in the cells of plants and animals and speculated that they must be or once have been bacteria. As early as 1893 the German biologist A. Schimper proposed the idea that the photosynthetic parts of plant cells came from cyanobacteria (or "blue-green algae" as they were called at the time). By the first quarter of this century, the American anatomist Ivan Wallin and the Russian biologist and scholar Konstantin S. Mereschovsky had independently come to the same conclusion. In 1910 Mereschovsky, who taught at the University of Kazan, published an essentially modern view of the origin of eukaryotic cells from various kinds of bacteria. It has taken the biological world over seventy years to catch up with him.

But the rise of modern genetics emphasized the nucleus of the eukaryotic cell, and by the twentieth century it was common to dismiss cell symbiosis theories as preposterous or absurd. So much was this the case that Professor Richard Klein of the University of Vermont could say as late as the early seventies that, "This bad penny [the theory of the symbiotic origin of plastids and mitochondria] has been circulating for a long time. . . . Clearly there is no chemical, structural, or phylogenetic basis for this belief."[30] As recently as two decades ago, college students were still being taught that organelles within eukaryotic cells probably "pinched off" from the nucleus and subsequently evolved their separate specific functions. With the ability to analyze DNA, however, it soon became clear that the pinch-off theory had a major failing: it couldn't explain the presence of DNA in certain organelles, nor the fact that this DNA was like that of bacterial genophores and different from that of the nucleus's chromosomes. In 1962 Hans Ris, a cell biologist at the University of Wisconsin, discovered structures looking just like the DNA in the chloroplasts of the green alga *Chlamydomonas.* He was immediately struck by the similarity in appearance of the DNA in the chloroplasts to that of cyanobacteria. From that time on "chemical, structural, and phylogenetic evidence" began to mount in support of the idea of a free-living bacterial origin for parts of nucleated cells.

Human religion and mythology have always been full of fantastic combinations of creatures—the mermaids, sphinxes, centaurs, devils, vampires, werewolves, and seraphs that combine animal parts to make imaginary beings. Truth being stranger than fiction, biology has refined the intuitively pleasing idea with its discovery of the overwhelming statistical probability of the reality of combined beings. We and all beings made of nucleated cells are probably composities, mergers of once different creatures. The human brain cells that conceived these creatures are themselves chimeras—no less fantastic mergers of several formerly independent kinds of prokaryotes that together coevolved.

• • •

No one has lived long enough to witness the origin of species in the field. But in one case in the laboratory, a new variety of microbe evolved so quickly it was caught in the act. This illustrative event was luckily witnessed and described by Kwang Jeon, a brilliant and keenly observant scientist in the zoology department of the University of Tennessee.[31] The symbiotic odyssey recorded by Jeon shows the dynamic we think responsible for the rapid evolution of cells with nuclei from anucleate bacteria about 1,500 million years ago. The story strongly demonstrates the inevitability of some kind of cooperation among organisms that are to live together and survive. It shows the thin line between evolutionary competition and cooperation. In the microcosm guests and prisoners can be the same thing, and the deadliest enemies can become indispensable to survival.

Jeon had been raising and experimenting with amoebae for years when he welcomed a new batch to his laboratory. After putting the new batch into special small bowls next to other amoebae gathered from all around the world, he noticed the spreading of a severe illness. Healthy amoebae grew round and granulated. They refused to eat and failed to divide. Bowl by bowl, more and more amoebae died. The few that grew and divided at all did so reluctantly, about once a month instead of once every other day.

When Jeon examined the dead and dying forms under the microscope, he noticed tiny spots inside the cells. On closer inspection he saw that about 100,000 rod-shaped bacteria, brought in by the new amoebae, were present in each amoeba. The rod bacteria had infected the rest of his collection. Yet the disease did not prove a total catastrophe. A small minority of infected amoebae survived the scourge. These "bacterized" amoebae were delicate, fragile organisms, oversensitive to heat, cold, and starvation. They were easily killed by antibiotics, which, while deadly to bacteria, did not harm his normal "nonbacterized" amoebae. A change was occurring. The two types of organisms, bacteria and amoebae, were becoming one.

For some five years, Jeon nurtured the infected amoebae back to health by selecting the tougher ones and letting the others die. Still infected, the amoebae began to divide again at the normal rate of once every other day. Reproductively speaking, they were as adapted as their uninfected ancestors. They were not rid of their bacteria—they all harbored "germs." But they were cured of their disease. Each recovered amoeba contained about 40,000 bacteria.

For their part, the bacteria had dramatically adjusted their destructive tendencies in order to live inside other living cells. Thus, from a violent confrontation emerged a new symbiotic organism, bacterized amoebae. Now, some fifteen years after the plague, the permanently infected amoebae are no longer sick but alive and well and living in Knoxville, Tennessee.

But the story does not end here. Applying his expertise in manipulating amoebae nuclei, Jeon followed up on his original experiments. From friends, Jeon reclaimed some of the amoebae that he had sent off before the epidemic and which had never been exposed to the pathogenic bacteria. With a hooked glass needle, he then removed the nuclei from both infected and uninfected organisms and exchanged them.

The infected amoebae with new nuclei lived on indefinitely. But the "clean" amoebae supplied with nuclei from cells that had been infected for years struggled for about four days and then died. It seemed as if the nuclei had become unable to cope with a "healthy" cell. Had they actually come to need their bacterial infection?

To find out, Jeon prepared another batch and mounted a rescue. Just a day or so before the bacterialess amoebae with their new nuclei would have died, he injected some of them with a few bacteria. The bacteria rapidly increased to the level of about 40,000 per cell, and the sick amoebae returned to health. A symbiotic habit had been formed; the bacteria were the "fix."

Jeon's amoebae can be killed by penicillin, which binds to sites in the cell walls of the bacteria within them, destroying the interdependent population that is the cell. The pact between bacteria and amoebae has become so intimate and strong that death to one member of the alliance spells death for both.

Jeon's amoebae showed that the only differences between organisms that kill or sicken, organisms that live together, and the indispensable components of organisms are differences of degree. Dangerous pathogens can become required organelles in less than a decade—very suddenly indeed considering the 350,000,000 decades or so of biological evolution. Symbiosis leads abruptly to new species. These new species—such as the bacterized amoebae—did not evolve gradually by accumulating mutations over a long period of time.

The amoebae experiments point out the fallacy of the idea that evolution works at all times for the "good of the individual." Just what is the "individual" after all? Is it the "single" amoeba with its internalized bacteria, or is it the "single" bacterium living in the cellular environment which is itself alive? Really the individual is something abstract, a category, a conception. And nature has a tendency to evolve that which is beyond any narrow category or conception.

One such conception is the popular idea that evolution is a bloody struggle in which only the strong survive. "Survival of the fittest," a motto coined by the philosopher Herbert Spencer (1820–1903), was used by late-nineteenth century entrepreneurs to justify such mean practices as child labor, slave wages, and brutal working conditions. Warped to mean that only the most ruthless win out in the "struggle for existence," it also implied that exploitation, since it was natural, was morally acceptable.

Darwin would have been shocked at the misuse of his ideas. He used Spencer's phrase "survival of the fittest" to refer not to large muscles, predatory habits, or the master's whip but to leaving more offspring. Fit, in evolution, means fecund. The point is not so much the infliction of death, which is inevitable, as the propagation of life, which is not.

Competition in which the strong wins has been given a good deal more press than cooperation. But certain superficially weak organisms have survived in the long run by being part of collectives, while the so-called strong ones, never learning the trick of cooperation, have been dumped onto the scrap heap of evolutionary extinction.

If symbiosis is as prevalent and important in the history of life as it seems to be, we must rethink biology from the beginning. Life on earth is not really a game in which some organisms beat others and win. It is what in the mathematical field of game theory is known as a nonzero-sum game. A zero-sum game is one like Ping-Pong or chess, where one player wins at the expense of his opponent's loss. An example of a *non*zero-sum game is children playing house, or war: more than one player can win, and more than one side can lose. Life is really far more a nonzero-sum game than many realize. Indeed, the political scientist Robert Axelrod recently convened a group of professional game theorists, economists, evolutionary biologists, and mathematicians from all over the world.[32] Convention participants submitted computer programs to play the nonzero-sum game *Prisoner's Dilemma*—a game in which both participants get three points if they both cooperate and one point if they both defect. If one player defects after the other has cooperated, the defector gets five points, while the cooperator receives none. It therefore seems at first glance that the best way to get points is to defect while you coax your partner into cooperating. But that is hard to do. Axelrod noticed that the most effective strategies were ones that were "nice," "forgiving," and "reciprocal." "Nice" meant not being the first to defect, "forgiving" meant not continuing to defect in further rounds to "teach the other player a lesson" if the other player had defected once but now stopped, and "reciprocal" meant cooperating (and defecting) at least once if the other player had done so. Greedy and bully programs that tried to take advantage by defecting with cooperators were consistently the worst performers. Then Axelrod undertook the exercise of "evolving" the computer programs by representing the highest scorers more fully in succeeding generations. After a sufficient number of generations, hardly any of the ruthless programs survived. Indeed, Axelrod concluded that, in a nonzero-sum game, cooperation increases with time. Axelrod's work is consistent with our view that all large organisms came from smaller prokaryotes that together won a victory for cooperation, for the art of mutual living.

It is not too whimsical to say that if we feel at loose ends, of two minds, beside ourselves, going to pieces, or not together, it is probably because we are. Real organisms are like cities: Los Angeles and Paris can be identified by their names, by their city limits, and by the general lifestyles of their inhabitants. But closer inspection reveals that the city itself is composed of immigrants from all over the globe, of neighborhoods, of criminals, philanthropists, alley cats, and pigeons.

Like cities, individual organisms are not Platonic forms with definite borders. They are cumulative beings with self-sufficient subsections and amorphous tendencies. And just as they are composites of species, they are also the working parts of larger superorganisms, the largest of which is the planetary patina. An organelle inside an amoeba within the intestinal tract of a mammal in the forest on this planet lives in a world within many worlds. Each provides its own frame of reference and its own reality.

Our own cells, like Jeon's amoebae, most likely require former bacterial intruders in order to live and breathe. Our ancestors have left genetic fingerprints in the cells that make us up. The history of the early, purely prokaryotic phase of life is preserved in particular by three types of intracellular structures: the mitochondria, the plastids (chloroplasts), and the undulipodia. Without them, neither our inner world, nor the outer world, nor the scanty interface of these two worlds we distinguish as human, would exist.

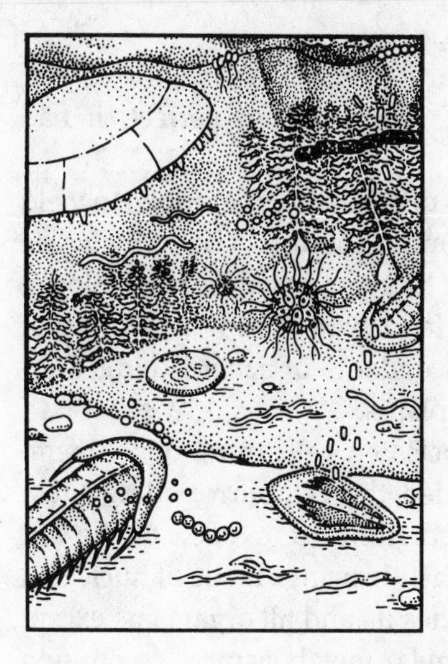

CHAPTER 8

Living Together

AS complex as it is compared to an inanimate droplet of chemicals, the prokaryotic cell is quite a simple affair among life forms. A membrane bounds some cytoplasm, dotted with hundreds of ribosomes, around the center of which floats a loose, unpackaged strand of nearly pure DNA, the repository of some 4,000 genes. The nucleated cell, by contrast, is larger, more complex, studded with mitochondria and plastids, and held together by a network of reticular structures and by the streaming, pulsing cytoplasm around them. The nuclear DNA, much of it repetitive, is coiled tightly into chromosomes which are contained in a membrane-bounded nucleus. With the keen perception of hindsight, eukaryotic cells now look like mergers of diverse organisms. Living corporations, some of the mergers began as hostile take-overs of one organism by another. But over hundreds of millions of years they became so coordinated, so interwoven that it took the electron microscope and the intricate techniques of biochemical analysis to penetrate the illusion that because the harmony of cell parts seemed perfect, it had always been so.

The first of the components of cellular corporations to yield clues to its origins was the mitochondria. Their small size and simple manner of dividing had convinced French biologist Paul Portier as early as 1918 (and Ivan Wallin by 1925) that mitochondria were direct descendants of bacteria lodged inside the cells of animals and plants.

Inhabiting nearly every nucleated cell, these dark, membrane-bounded bodies provided the cell surrounding them with abundant energy derived from the oxygen of the air. Because of mitochondria, all Earthly beings made of nucleated cells—which, of course, includes us and all organisms except bacteria—have remarkably similar metabolisms. Discounting the photosynthetic metabolism monopolized by plants and algae (which is virtually identical to that of cyanobacteria), in all its fundamental details eukaryote metabolism is the same. Bacteria, by contrast, exhibit a far wider range of metabolic variations than eukaryotes. They indulge in bizarre fermentations, produce methane gas, "eat" nitrogen gas right out of the air, derive energy from globules of sulfur, precipitate iron and manganese while breathing, combust hydrogen using oxygen to make water, grow in boiling water and in salt brine, store energy by use of the purple pigment rhodopsin, and so forth. As a group bacteria obtain their food and energy by ingenious methods, using every sort of plant fiber and animal waste as a starting material. (If they did not, we would be living in a mounting heap of garbage.) We, however, use just one of their many metabolic designs for energy production, namely that of aerobic respiration, the specialty of mitochondria.

In eukaryotic cells wastes from fermented food molecules in the cytoplasm such as alcohol and lactic acid enter the mitochondria and pass through a cycle of reactions involving oxygen and the same sorts of electron chains that are found in aerobic bacteria. These reactions inside the mitochondria produce most of the ATP for both the mitochondria and the rest of the cell. The oxidation of food molecules yields carbon dioxide and water as waste.

Mitochondria have retained many intriguing signs of their former free-living status. Although they lie outside the cell's nucleus, they have their own genetic apparatus, including their own DNA, messenger RNA, transfer RNA, and ribosomes enclosed in mitochondrial membranes. Like bacterial DNA, their DNA is not bound into chromosomes, and it remains uncovered by the histone (a type of protein) coating found in the nuclear DNA of their home cell. Mitochondria assemble proteins on ribosomes which are very similar to the ribosomes of bacteria. Moreover, mitochondrial ribosomes and those of respiring bacteria tend to be sensitive to exactly the same antibiotics, such as streptomycin.

Like most bacteria and unlike the complicated reproduction of the rest of the nucleated cell, mitochondria pinch and divide in two to reproduce, usually doing so at different times from each other and the rest of the cell. Studies at the Carlsberg Laboratory in Copenhagen and at the French National Center of Scientific Research Laboratory at Gif-sur-Yvette have shown that mitochondria engage in the unsystematic genetic transfer that characterizes bacterial sex.

These and other signs point to the explanation that mitochondria were once bacteria that became symbiotically holed up inside larger bacterial cells. In damp Proterozoic soil or in a microbial mat with grass-green and blue-green bacteria bubbling maddening amounts of polluting oxygen from their every surface, forcing all around them to evade it or evolve, a type of oxygen-breathing bacterium arose. It was probably a fierce predator, perhaps resembling modern predatory bacteria such as *Bdellovibrio* and *Daptobacter*.

Bdellovibrio are oxygen breathers that burst asunder their bacterial prey, eating them out from the inside. Derived from the Greek, their name describes their wicked method: *bdello* means leech, and *vibrio* refers to their vibrating comma shape. *Bdellovibrio* attack by attaching to their prey and rotating like whirling drills to enter their victim's innards and break down the genetic material. After they have used it to make their own genes and proteins, they rupture the now useless and empty cellular bag of their ruined hosts.

Daptobacter, another moneran predator, also viciously attacks bacteria. In 1983 Dr. Isabel Esteve at the Universidad Autónoma de Barcelona in Spain realized that unlike *Bdellovibrio, Daptobacter* (the "gnawing bacterium") enters both the inner and the outer membranes of its victim's cell walls. Then it neatly divides, again and again and again, under either aerobic or anaerobic conditions. Imagine the ancestor of our mitochondria: a ruthless attacker, capable of breathing oxygen when it was around, or maybe even doing without it if necessary. The ancestors of mitochondria invaded and reproduced within our other bacterial ancestors. At first the occupied hosts just barely kept alive. But when they died, they took the invaders with them. Eventually only cooperators were left. The invaded victims and tamed mitochondria recovered from the vicious attack and have lived ever since, for 1,000 million years, in dynamic alliance.

In the long run, the most vicious predators, like the most dread disease-causing microbes, bring about their own ruin by killing their victims. Restrained predation—the attack that doesn't quite kill or does kill only slowly—is a recurring theme in evolution. The predatory precursors of mitochondria invaded and exploited their hosts, but the prey resisted. Forced to be content with an expendable part of the prey (its waste) instead of the entire body of the prey, some mitochondria precursors grew but never killed their providers. Time wreaked its changes on both parties. Animosity became interchange. Philip John of the University of Reading believes some cancers represent a sort of atavistic return to the original state of prokaryote animosity. From the peculiar behavior of mitochondria in many cancerous tissues John has concluded that mitochondrial rebellions have not been permanently quelled in all cases.

Eventually some of the prey evolved a tolerance for their aerobic predators, which then remained alive and well in the food-rich interior of the host. The two types of organisms used the products of each other's metabolisms. As they reproduced inside the invaded cells without causing harm, the predators gave up their independent ways and moved in for good. Like the tamed boars living peacefully in the barnyard as pigs, or "man's best friend," the dog which was once a wolf, deadly "disease" germs were domesticated, rendered harmless. According to a popular hypothesis for the development of the nuclear membrane, the prey was obligated to its oxygen-using guests because they protected its DNA from the poisonous oxygen increasingly prevalent in the outside world.

The original prey was probably a larger bacterium like *Thermoplasma*. Though it tolerates oxygen, *Thermoplasma* can use it only in small quantities, far smaller than those present in our modern atmosphere.[33] The prey bacterium may have been a tough microbe, able to protect the mitochondria from harsh environments. Surviving very hot and acidic waters such as those found in the hot springs of Yellowstone National Park, *Thermoplasma* is so far the best guess for the bacterial ancestor of the main, nucleocytoplasmic portion of the eukaryotic cell. And there is another reason for believing that *Thermoplasma* is the best candidate for the ancestral host microbes. Its DNA, unlike that of other bacteria, is wrapped in certain proteins similar to the histones of nearly all eukaryotes.

Steroid synthesis may have been an emergent property of the mitochondria symbiosis. To make flexible membranes, eukaryotes pack them with steroids, complex four-ring lipid molecules that require atmospheric oxygen during their fabrication. Steroids "lubricate" the membrane proteins, making the internal and external membranes of eukaryotes loose, easy to break and fuse over a wide range of temperatures. The steroid-containing membranes form vesicles. They wrap up the mitochondria as well as the nucleus itself. The biochemical steps involved in the synthesis of steroids begin in the cell cytoplasm, but the last steps in the process occur in the mitochondria. Because it needs oxygen and seems associated with the mitochondria, steroid synthesis may have arisen from the new relationship between aerobic bacteria and their unwilling hosts, between predator and prey.

As the predator made itself at home, it gradually lost some of its DNA and RNA. Natural selection tends to eliminate redundancy when symbioses evolve, so if both organisms synthesize a required nutrient, for example, one of them may gradually lose the ability to do so, thereby increasing their interdependency. Now, mitochondria are totally dependent on the rest of the cell. They share the host cell's genes which code for the production of some of their proteins, including some of the enzymes required for the replication of their own DNA and RNA. The cell uses the energy that mitochondria derive from oxygen, and the mitochondria use the organic acids which were the prey cell's waste materials. When these processes cease, we and all other composite beings die.[34]

And the prey? The larger cells invaded by the parasites must have been bacteria, too. They were probably similar to the modern *Thermoplasma*, or to its relative *Sulfolobus*. A tough, versatile bacterium, *Thermoplasma*'s ability to withstand extremes of temperature and acidity was a priceless attribute in the Proterozoic world. *Thermoplasma* and *Sulfolobus* belong to a newly recognized group of bacteria with distinctive ribosomes and ribosomal RNA. Such organisms are called archaeobacteria (old bacteria) because their sharing of traits with each other and all eukaryotes suggests to some scientists an ancient divergence from the rest of the bacteria. Other, more familiar bacteria—like *E. coli* and *Bacillus*—are known as eubacteria (true bacteria).

Cohabiting symbiotically, the archaeobacteria and their eubacterial invaders did what neither could do alone. Their descendants became the foundation of the macrocosm. All the familiar creatures of the Earth today, from seaweed to sea urchin to sea lion to sailor, are composites of nucleated cells. The nucleated cells themselves are the result of prokaryotic mergers. And each cell with a nucleus is packed with the deep-breathing mitochondria that once upon a time were bacteria.

Maybe a hundred million years after mitochondria had become established, a new type of organism joined them in the cytoplasm of certain cells. But the genesis of the union was not through infection but eating. Like Jonah swallowed by the whale, the forebears of plastids—the photosynthetic parts of nucleated cells—were probably eaten by hungry protoeukaryotes. These ancestors were probably the ubiquitous cyanobacteria, the same successful beings which wove the microbial mats and built the stromatolitic skyscrapers. Routinely eaten, some of these bacteria apparently resisted digestion inside their hosts, their valuable light-trapping pigments still active.

Today, locked inside every plant and many protists (such as the mobile *Euglena*), plastids ply the biosphere with food and oxygen—a far greater contribution than any made by the world's entire population of mammals. Plastids make food from water and sunlight. Mammals—including humans, of course—do no such thing. From a planetary point of view, the major role of mammals may be as fertilizers of plants and carriers of mitochondria. But if all mammals were to die in one instant, insects, birds, and other organisms would carry mitochondria and fertilize plants. If plants with their plastids were to suddenly disappear, however, the output of food on the planet would be so severely hampered that all mammals would certainly die.

The green plastids—called chloroplasts—in plants and green seaweeds are larger and even more conspicuously similar to bacteria than are the mitochondria. Plastids also have their own DNA and messenger RNA. Their ribosomes, too, are the same size as those in bacteria. Like mitochondria, plastids are wrapped by a membrane which separates them from the rest of the cell. Like bacterial DNA, plastid DNA lacks the histone tangled up with the DNA of chromosomes. Plastids inside cells divide directly in two, like bacteria. And the DNA, RNA, and proteins produced by plastids are strikingly similar to those of bacteria, especially cyanobacteria.

Despite having shed most of their tools of self-sufficiency, plastids can synthesize many of their own proteins. Most, but not all, have retained the pigmented green, red, or blue-green photosynthetic systems which earned them their keep. While all plant cells contain some sort of plastid, some colorless ones are nonphotosynthetic. No one knows exactly what function the pale plastids serve, but their presence still bespeaks an ancient alliance of disparate organisms.

In the late 1960s, Ralph Lewin, a marine biologist at Scripps Institute of Oceanography in California, discovered an obscure grass-green bacterium in various locations within the sun belt of the tropical Pacific Ocean—off Baja California, Singapore, and the Palau Islands. He called the microbe *Prochloron. Prochloron* grows on marine animals known as sea squirts. It coats the quiescent, lemon-shaped organisms, coloring them green and possibly supplying them with certain nutrients. Some sea squirt larvae even carry pouches full of *Prochloron*, insuring their symbiotic growth with the next generation of animals. The large cell size and grass-green color of *Prochloron* had led to the assumption that, like the protist *Chlamydomonas*, or the symbionts of green hydras, *Prochloron* was a green alga. The first electron micrographs, however, proved to be quite a shock. *Prochloron* was not an alga at all but an enormous bacterium.

Little more than a DNA machine bounded by a membrane and loaded with bright pigments, *Prochloron* would be a chloroplast if it did not have a bacterial cell wall and if it lived inside a plant cell instead of on the outside of a sea squirt. *Prochloron*—a missing link in the history of symbiosis—combines the physiology of a plant with the structure of a bacterium. (Cyanobacteria are also green—actually blue-green—but their physiology is distinct from that of green algae and plants. *Prochloron* contains both chlorophyll "a" and chlorophyll "b," making it significantly more like plants from a metabolic standpoint than are cyanobacteria, which contain only bluish-green pigments and chlorophyll "a.") It seems very possible that the ancestors of *Prochloron* were eaten by many kinds of protists. Some resisted digestion and those that stayed alive eventually evolved into plastids. Today plants turn toward the sunlight because, without it, their fussy guests would die.

W. Ford Doolittle, a biochemist at Dalhousie University in Nova Scotia, compared the sequence of nucleotide bases in the ribosomal RNA of red plastids in the seaweed *Porphyrid-*

ium with that of the RNA in the seaweed's own cytoplasm. He found less than 15 percent similarity. He then compared the sequence with the ribosomal RNA in the cyanobacterium *Synechococcus* and with that in a green plastid from the protist *Euglena*. The similarities measured 42 percent and 33 percent respectively. This convinced him and others that cyanobacteria are ancestral to the red plastids, or rhodoplasts, in red algae. In like fashion—although the detailed molecular biology has still not yet been done—the ordinary green plastids of plants probably come from *Prochloron*.

As you look around at the world of nature you cannot fail to see the tremendous success of *Prochloron*'s descendants: jungles, gardens, house plants, and grassy hills, all of them testify to the success of plastids. Eaten, but not digested, they have insinuated themselves into every corner of the world, hitchhiking as part of a cooperative partnership called the eukaryotic cell.

The new eukaryotes had diversified: some now had two basic ways of generating ATP—respiration and photosynthesis. Those that were also able to swim were virtually unstoppable as they took over compatible environments farther and farther afield. It is no wonder that in forms such as algae and phytoplankton, they began to dominate the oceans and other wet places of the world. Indeed, they continued to evolve to the point that they left the water, took over the land, and eventually became the primeval plants of the macrocosm.

CHAPTER 9

The Symbiotic Brain

FROM our present distance in evolutionary time we cannot say exactly when, but around or slightly before the time bacterial mergers gave cells the ability to use oxygen and light, they also seem to have conferred on life another capacity: that of motility. By joining the big, new cells, rapidly moving bacteria gave them the basic advantages of locomotion—avoiding danger and seeking food and shelter. Other benefits of travel—a greater selection of habitats, more opportunities for genetic exchange—came within reach. Mobility, however, was only the most obvious benefit of these partnerships.

If you look at a living eukaryotic cell under the microscope, you may be startled by the vigorous movement within it. In sharp contrast to a bacterial cell whose contents are motionless or drift passively about, the interior of eukaryotic cells is swarming like a city. The cytoplasm streams. In some cells mitochondria, ribosomes, and other organelles course about on predetermined tracks as though obeying two-lane traffic signals. Many cells rhythmically expand and contract. In a chameleon that is changing color, particles of pigment are carried from the surface to the interior of cells when the skin of the animal lightens. In actinopods, a kind of protist (see note 1), long moving spikes used to grab prey or as stilts for "walking," extend out from the cell surface. Most such cell motion takes place along an elaborate transportation system within the cell made of microtubules—tiny tubes of protein only 240 angstroms in diameter which, in concert with other proteins, can cause motion.[35] Microtubules are visible only with an electron microscope.

We believe that the nucleated cell's ability to move both without and within is the contribution of another symbiotic merger with bacteria, this time with rapid, whiplashing spirochetes. Unlike the similar theories for mitochondria and plastids, this hypothesis is still not popular among biologists. No DNA has turned up among the pertinent apparatus within the cell to spell out the code of a once-foreign cell government. Without it, many scientists still are reserving judgment on the issue of a bacterial origin of cell projections made of microtubules. But RNA, which has been located in some of the parts in question, may itself be responsible for the construction and replication of motility structures without direct help from DNA. As we saw earlier in the chapters on preanimate matter, RNA is capable of replicating itself, and may have evolved even earlier DNA. If spirochete RNA and proteins are found that match RNA and proteins in the cell's motion structures more closely than they do any randomly chosen RNA and protein, the spirochete symbiosis hypothesis will become hard to oppose.

Of course, there is a paradoxical inverse relationship between symbiosis and evidence. Partner organisms that live together in near-perfect harmony will be almost indetectible. Oxford University Botanist David Smith compares what remains from such mergers to the smile of the Cheshire Cat, the feline character in *Alice in Wonderland* that slowly disappeared until all that was left was an enigmatic grin: "the organism progressively loses pieces of itself, slowly blending into the general background, its former existence betrayed by some relic."[36] In this light, certain telltale signs become all the more provocative.

Close study of undulipodia—tiny cell whips on many kinds of cells with nuclei—shows an amazing uniformity of structure across a vast array of organisms. These supple filaments have traditionally been called flagella[37] if they are long and few like sperm tails, or cilia if they are short and many like hairs, but there is no basic difference between them. Nearly all algae, ciliates, and slime molds—that is to say, protists, the earliest organisms with nucleated cells to have evolved—have them. Their whipping or waving motion propels a free-swimming cell through its medium or, if the cell is fixed in place, moves particles past it. From the sperm of a male fern to the lining of a mouse's nostrils, many cells of complex organisms sport these undulating organelles.

No matter what cell or organism they adorn, undulipodia are all about a quarter of a micron in diameter, and in cross section show a telephone dial-like circle of nine pairs of minute microtubules surrounding another pair in the center. This pattern, known as the 9 + 2 array (strangely, since "nine" refers to the pairs in the outside circle while the "two" refers to each individual tubule of the center), is found in the sperm tails of bulls, of whales, and of ginkgo trees; in the cilia of our lungs, windpipes, and oviducts and those of other mammals; in the antennules of lobsters; in the cilia that cover that familiar protist, the paramecium; and in those on the zoospores of water molds. Indeed, this is just a tiny sampling of the prevalence of these 9 + 2 patterns.

Further, an undulipodium invariably grows out of a structure called a kinetosome which is composed of nine triplets of microtubules arranged in a circle. The walls of all these microtubules contain two related proteins, alpha and beta tubulins. Nearly two hundred proteins, including an extremely complex one called dynein, make up the rest of the 9 + 2 structure. Many of the nontubulin proteins have not yet been isolated or named, but the combined evidence is so conclusive that all evolutionary biologists believe these 9 + 2 structures could not have evolved separately in protists, plants, and animals. (Since fungi, although they always lack undulipodia, do have their component microtubules and tubulin proteins, it seems likely that their ancestors had cell whips but lost them.) There was, rather, a common origin.

Our candidate for this common ancestor is the spiralling, motile, hairlike spirochete, the fastest bacterium in the microcosm. In the sticky regions of their microworld of gelatinous muds and viscous fluids, spirochetes are often the only bacteria capable of passing through a certain region. The spirochete's *métier* is motion. Some spirochetes even contain microtubules, although none, so far, have been found arranged in a telephone-dial array. To us the evidence strongly suggests that ancient pacts were made between the early bacterial confederacies that became cells with nuclei and spirochetes or spirochetelike bacteria. Spirochetes hovered both inside and outside their nonspirochete neighbors, and in the end they provided efficient movement for those who had never even requested it. Of course this is just a hypothesis, but it aids in the explanation of otherwise disparate data. As Darwin said in 1859 in *The Origin of Species* regarding his theory of natural selection: "Anyone whose disposition leads him to attach more weight to unexplained difficulties than to the explanation of a certain number of facts will certainly reject my theory."[38] The same holds true for the spirochete hypothesis. It may seem bizarre, but it explains so many facts that the story of the microcosm, of the evolution of sex and meiosis in the so-called higher organisms, suffers without it.

As usual, hunger was probably the impetus for these new mergers. Our protist ancestors were often hungry and sometimes tortured by starvation. Each bacterial symbiont in the new collective required food, and immobile alliances were at the mercy of their environment. During periods of dryness and scarcity, the new cells could do nothing but wait. Some met environmental threats by forming resistant, thick-walled cysts, as many cells do today. As these ancient, unicelled ancestors stagnated, many different moving bacteria bumped into and bounced off them in their own search for food. Some of these swarming, whiplike bacteria thrived around the cysts, grooming their surfaces, skimming the fat and collecting the crumbs. Some devious, infiltrating spirochetes were probably pathogenic to the bacteria with which they came into contact. Others must have been totally harmless.

Free-living, scavenging spirochetes are still well known today, as are many varieties engaged in symbiotic or parasitic lifestyles with other organisms, such as insects, mollusks, and mammals, humans included. Some live harmlessly in our gums; others, like *Treponema pallidum*, are found in the blood of people suffering from syphilis.

Spirochetes tend to attach to things, living or not. When they swim next to each other, they also tend to undulate in unison simply due to their proximity in a liquid medium. As scavenger spirochetes feed on the surface of their host, particularly if they are amassed together on one side, they can propel it through its medium with their coordinated undulations.

Those spirochetes and protists that coevolved elegant attachments swam well. Consequently they found more food and reproduced more often—a clear advantage. Natural selection would undoubtedly have favored these alliances until the two partners gradually became one. A certain modern amoeba, for example, draws in its undulipodium and gorges itself when food is plentiful, moving about slowly and dividing in a typically sluggish amoebic manner. When food becomes scarce, however, it grows a kinetosome and an undulipodium and swims away in search of food.

The advent of spirochete alliances 2,000 million years ago must have altered the microcosm, as the steam engine altered human civilization. The new motile eukaryotes must have revolutionized the bacterial world by their sudden boost to microbial transportation and communication. The traffic of cells and thus the flow of genetic information accelerated. And as the steam engine sped up the cycles of industrial production, including the manufacture of more steam engines, spirochete partnerships may have initiated a burst of development, increasing the number and diversity of symbiotic life forms.

Certainly the development of good attachments for symbiotic spirochetes probably caused a burst of microbial speciation. There exist today about 8,000 different species of free-living, unicellular protists called ciliates. Each ciliate species is recognized by the characteristic number and grouping of patterns of cilia on the surfaces of its members. Indeed, the patterns of cilia on the surfaces of some microbes can be inherited independently of the cell's nuclear genes. If in microsurgery an area of the surface is removed and replaced upside down, this upside-down pattern will be directly inherited by offspring cells after cell division. This, too, implies an independent origin for the undulipodia of cells.

Modern-day spirochetes still readily enter symbioses for

the purpose of mobility. In the hindgut of a certain voracious Australian termite lives the protist, *Mixotricha paradoxa*, which appears to be covered with cilia. Only four of them at the forward end, however, are genuine 9 + 2 undulipodia, and they take no part in moving the creature. The rest of the "cilia" are actually 500,000 *Treponema* spirochete bacteria, attached to its surface, feeding together and waving in unison. The four small undulipodia merely serve as rudders, permitting *Mixotricha paradoxa* to change direction.

Termites teem with symbiotic microbes. Without them, the wood-eating insects would starve to death because they are unable to digest cellulose. Yet termites acquire their vital supply of bacteria not through their genes, but in a peculiar ritual of feeding on the anal fluid of their fellow termites. The spirochetes, swimming alone and en masse, propel particles of food, each other, and larger microbes about within the termite's swollen intestine, called the hindgut. Oblivious to the great world of larger organisms outside the termite, the spirochetes enjoy a rich food supply and protection from exposure to oxygen.

How cells got the spirochetal microtubules within their walls and built from them a brilliant apparatus for organized movement is one of nature's most perplexing puzzles. Volumes have been written about the intracellular microtubular system itself. And for good reason; microtubules are involved in cell secretion, cell division, and nerve cell formation. Microtubules are therefore part of the story of cancer and brain development. Most of the thousands of people who work on microtubules in nerves, brains, sperm, and spiny protists, isolating and studying their tubulin proteins, never stop to ask where microtubules come from in an evolutionary sense. If confronted with the case for a spirochetal origin, they might reply, as did Michael Sleigh, a professor of biology in Southampton, England, "Microtubule systems were probably developed to support cell projections and direct and perhaps drive vacuolar circulation. . . . Such an origin . . . seems more likely than a symbiotic origin from spirochetes . . ."[39] But such an argument ignores the basic tenet that evolution never plans in advance. Because cells need microtubules and undulipodia does not mean they evolve them. Hungry squirming symbionts, on the other hand, attaching to cells for their own purposes, could have become the cell whip system without planning ahead. They could do so by obeying the same old story lines of vicious attack, compromise, and ultimate partnership of victor and vanquished. In their frantic search for food spirochetes failed to devour their victims and gave them rapid movement instead. But that spirochetal remnants compose the microtubule system of all eukaryotes and that captive spirochetal behavior is the common denominator of many elaborate functions is now only a tantalizing hypothesis.

The most intriguing microtubular performance might well be the intricate "dance of the chromosomes" known as mitosis, for that method of cell division led, among other places, to the animal cell—to *our* cells. It is this intracellular ritual, more complex than any ballet, that transforms one into two. The chromosomes of a cell about to divide double themselves. Each pair is joined somewhere along its length by a small disk called a *kinetochore,* the hook or button that connects chromosomes to microtubules. Two poles, like the north and south poles of the Earth, are marked in animal cells by small bodies called *centrioles.* An array of fibers mysteriously fans out between the poles forming a spindle. At a closer look these fibers are seen to be hundreds of microtubules. The doubled chromosomes line up along the cell's equator, each fastened to a fiber by a kinetochore. The kinetochores then duplicate, freeing each member of the chromosome pair. As long as they are connected to the microtubules, the chromosomes separate, moving to opposite poles. A membrane forms around each group of chromosomes, making two new nuclei, and the cell subsequently divides in two.

This amazing, stately pavanne ensures the precise replication and division of genetic material to the offspring cells—quite a feat considering cells with nuclei generally have thousands of times more DNA than bacteria. As we have seen, all the DNA of all your own cells, stretched out end-to-end, would reach from the surface of the Earth to the moon and back over a million times; and yet this DNA, especially when it has to move during the division of the cell, stays neatly packaged in your chromosomes. In imagining how this process came to be, it helps to bear in mind that selection pressures must have been relentless: cells lacking genes, and those with incomplete sets of DNA, died.

But the intricacy of the mitotic dance becomes still more understandable if you allow for a spirochetal choreographer. And the clues are there. The spindle is made of microtubules, the same microtubules found in all cell whips. The centrioles at the cell poles are composed of the same telephone-dial structures of nine triplet microtubules arranged in a circle. Indeed, centrioles are identical to kinetosomes, the only difference being the technical distinction that centrioles do not have cell whips. Yet in many organisms, the centrioles can rapidly transform into kinetosomes simply by sprouting cell whips—undulipodia—as soon as cell division is over.

The timing of cell processes provides a clue to the varied roles spirochetes played in evolution. In plants and animals, undulipodia and mitosis are mutually exclusive—they are never seen in the same cell at the same time. Fungal cells seem to have permanently traded cell whips for mitosis. But for some protists to divide, they must first pull their undulipodia inside their cells. No mammal cell—not to mention many other kinds of cells—can retain undulipodia while it divides by mitosis. It is as if the cell must use its ancient spirochetal symbionts for one thing or the other, but not both.

Like comic-book characters trying to discover the true identity of a masked superhero, scientists have noticed that the two microtubular structures are never in the same "room" together, but haven't yet been able to prove that they are one and the same being. We cannot prove that the mitotic spindle and the waving cell whips are the same spirochetal remnants in disguise. But the data suggests a single biological identity.

Unfortunately, the scientist cannot see the spirochete's wriggling bodies embedded and directly dividing as organelles in the cell cytoplasm, like chloroplasts and mitochondria. Instead, he or she sees only what are known as microtubule organizing systems, tiny centers viewed with the electron microscope that precede the growth of single microtubules, centrioles, kinetosomes, and actinopod spikes. Many biologists have asked why the cell "bothers" to make the elaborate telephone dials, and they have called centrioles a "central enigma" in biology.[40] Yet if the theory of symbiotic spirochetes is right, all 9 + 2 structures are evolutionary relics; and the microtubule organizing centers are the actual remnants of the spirochete populations that were once adopted by host cells.

When microbial communities becoming eukaryotes first acquired nucleic acid from spirochetal invaders, it was probably used only to direct the production of more offspring spirochetes to propel the community. But the invaded communities of bacteria that became new cells then adopted the spirochete genes to coordinate the division of the cumbersome genetic material of the community. The result of such adoption of genes inside new cells was often an absence of undulipodia and motility outside the cell.

Today, all fungi and red seaweeds, many amoebae, green algae and many other eukaryotes have kept mitosis but not undulipodia; such beings show both the desirability of a sophisticated internal system for the transportation of chromosomes during reproduction, and the problem with having such a system and external undulipodia at the same time. These organisms never solved the essential contradiction of keeping both internal chromosome-moving equipment and protruding cell whips. Both were based on the microtubule system. Chromosome movement in all reproduction was more important, so they lost their cell whips.

How to retain mitotic division without sacrificing cell whip motion was a dilemma that yielded a vast array of experimental solutions among life forms. A good many were successful enough to have persisted until today. Some organisms, for example, had both mitosis and motility, but at different stages in their life cycles. The ciliates retained hundreds of cell whips (cilia) by dividing their nuclei by methods other than standard mitosis. Still others, some of which ultimately gave rise to plants, only form undulipodia during the development of their sperm. Though motile, the sperm tails are undulipodia and so these cells can never divide by mitosis. They are destined only to fertilize eggs which, although they can divide, never have undulipodia.

Probably thousands of variations on these themes emerged and vanished. Such is the way of nature that solutions so perfect they left no room for further improvement often came to a premature end. Organisms that managed to retain motility and divisibility in a single cell for the most part remained to this day single cells.

The step that was to be so spectacularly successful was the one that kept the most options open: organisms arose that grew an undulipodium in one cell, but reserved another attached cell that could still divide. The undulipodium could then propel both cells.

This innovation opened the way for new heights of structural complexity. Evolving from it were organisms that have a germ cell and a body—a germ cell for the sole purpose of reproduction, and a multicelled body highly differentiated for myriad specific functions. This, indeed, is the basic plan of our bodies. Our germ cells, sperm and egg, are the only cells that can reproduce a human being, even though each of our body cells may contain a complete copy of all our genes.

The body is totalitarian in its regulation of genes. Once a cell becomes a muscle cell, for example, it is so forever. The only exception to this rule of permanent roles within the body is during cancer, when cells seem to revert back to the more primordial condition of reproducing continuously without regard to their place or function in the body. During cancer, chromosomes break apart and mitochondria reproduce even more rapidly than the cells of which they are a part. Usually once a cell commits itself to growing an undulipodium it is evolutionarily dead: it cannot grow again. But as if disobeying all authority, some cancer cells in tissue culture even grow undulipodia, which they withdraw just before mitosis. It is as if the uneasy alliances of the symbiotic partnerships that maintain the cells disintegrate. The symbionts fall out of line, once again asserting their independent tendencies, reliving their ancient past. The reasons, of course, are not all that clear, but cancer seems more an untimely regression than a disease. Genes are regulated and cells differentiated in the body by the complex interaction of biochemicals within the body. When these biochemicals are diluted by the introduction of cigarette smoke, sodium nitrate, and other carcinogens, they cannot perform their task. Consequently cells tend to behave like children in a classroom whose teacher has left: they go wild, they get out of their cellular "seats," they play and reproduce in an unregulated, wanton fashion.

It is easy to forget that symbiosis is still going on throughout the biota. The spirochetes, for example, are still trying to make a living off their partners and victims. Symbiotic spirochetes act like undulipodia and true undulipodia act like independent organisms. Severed undulipodia of sperm cells, that is, sperm tails, broken off from the main cell body, swim away and survive for minutes or hours. On the other hand, healthy spirochetes may penetrate the wall of protist cells and enter their inner sanctuary, where they continue to swim around. Sometimes, inside the cell on their enigmatic mission, they reproduce. Just what they are doing is not clear, unless it is showing the human observers, themselves survivors of bacterial matches and mismatches, how history repeats itself.

Or perhaps we should say that history spirals back on itself. For within the eye that peers through the microscope, tiny rods and cones—nerve cells specialized for light perception—respond to the light and to each other by sending chemical and electric messages along axons and dendrites—the fibrous arms of neurons—to the brain. Cross sections of the rod and cone cells in the eye reveal the 9 + 2 pattern of microtubules. The axons and dendrites of the brain are a differently organized mass of microtubules, containing all the microtubular proteins but without the 9 + 2 formation. Something in the eyes triggers waves of transmissions across the synapses between the densely packed axons and dendrites of brain cells. Riding these waves is the thought: "Did the spirochete motility system of the microcosm evolve within the ordered environment of larger organisms to become the basis of their nervous systems?"

Proof of spirochete identity in the cells of the brain, beyond the rich presence in them of microtubules (neurotubules), is slowly accruing. Alpha and beta tubulins are the most abundant soluble proteins in the brain. Two or three proteins in termite-dwelling spirochetes have immunological similarities to tubulins in the brain and in all undulipodia. After maturity, brain cells never divide, nor do they move about. Yet we know mammal brain cells—the richest source of tubulin protein anywhere—do not waste their rich microtubular heritage. Rather, the sole function of mature brain cells, once reproduced or deployed, is to send signals and receive them, as if the microtubules once used for cell-whip and chromosomal movement had been usurped for the function of thought.

In explaining his thought process in a letter to Jacques Hadamard, Albert Einstein wrote that to explore the physical world he used nonverbal symbols, an abstract language of "reproducible elements." Explaining how he came to his spectacular discoveries, Einstein said: "The words or the language, as they are written or spoken, do not seem to play any role in my mechanism of thought. The psychical entities which seem to serve as elements in thought are certain signs and more or less clear images which can be 'voluntarily' reproduced and combined." Einstein went on to assert that to him there was no essential difference between the "mere associating or combining of reproducible elements and understanding [itself]."[41] If spirochetes are truly ancestral to brain cells or neurons, then the concepts and signals of thought are based on chemical and physical abilities already latent in bacteria. The "reproducible elements" of Einstein's brain may have preceded him in a different form, flourishing three billion years ago in the anaerobic environment of the early earth.

Einstein believed that his thought process proceeded by a "rather vague play with the above mentioned elements."

Taken from a psychological viewpoint, he sums up: ". . . this combinatory play seems to be the essential feature in productive thought—before there is any connection with logical construction in words or other kinds of signs which can be communicated to others." In a similar vein, the brilliant mathematician John von Neumann contrasted pulsing nerve cells of human cognition with the operations of computers. Since nerve cells affect only their immediate neighbors and then with relative slowness and imprecision, von Neumann speculated that they made up for their organic sluggishness by being far greater in number and acting simultaneously (in parallel) rather than in the serial, one-calculation-after-another fashion of all presently existing computers. (The Japanese, however, are furiously attempting to develop so-called fifth-generation, parallel computers that more nearly mimic the human brain.) In addition, von Neumann suspected that even though they were at the basis of digital computers, the four basic "species of arithmetic," that is, addition, subtraction, multiplication, and division, were not the true mathematics of the nervous system. Just as languages like Greek or Sanskrit are historical facts and not absolute logical necessities, von Neumann had come to believe that logics and mathematics were similarly historical, accidental forms of expression. Near the end of the last book he ever wrote, *The Computer and the Brain,* he recorded that "logics and mathematics in the central nervous system, when viewed as languages, must structurally be essentially different from those languages to which our common experience refers."[42] Could the true language of the nervous system then be spirochetal remnants, a combination of autocatalyzing RNA and tubulin proteins symbiotically integrated in the network of hormones, neurohormones, cells, and their wastes we call the human body? Is individual thought itself superorganismic, a collective phenomenon?

Again and again, study of the microcosm brings home to us that human capacities grow directly from other phenomena. Nature has a certain subsuming wisdom; our aptitudes must always remain meager in comparison to the biosphere of which we form relatively tiny parts. But we are not discontinuous from the general path of evolution, from the flows and flux of matter, information, and energies. Nor can human thought—the last refuge of those insisting on human "higherness"—be isolated or dissociated from the prior accomplishments of life. All our favorite inventions were anticipated by our planetmates; why not thought? If bacterial "cold light" (bioluminescence) preceded electric lights by 2,000 million years, if the protist *Sticholonche* propelled itself with microtubular oars through the Mediterranean long before Roman galleys rowed the same waters, if horses roamed the plains, whales the seas, and birds the skies long before the traffic of cars, submarines, and airplanes, is it so farfetched after all that bacterial symbionts created biospheric information pathways as important as quantum mechanics or the theory of relativity? In a sense we are "above" bacteria, because, though composed of them, our power of thought seems to represent more than the sum of its microbial parts. Yet in a sense we are also "below" them. As tiny parts of a huge biosphere whose essense is basically bacterial, we—with other life forms—must add up to a sort of symbiotic brain which it is beyond our capacity to comprehend or truly represent.

Our own human history spans leaps of advancement in our stretch of the world that are nearly as astounding as the leap from bacteria to the brain. Our biology produced our technology, which in turn catapulted us far beyond what we thought were our biological limits. For example, while early humans had only their own bodies as a medium for long distance communication—running, yelling, drumming—

we have accelerated from horsepower, automobiles, radio waves, electricity, and aerodynamics to sending our messages via satellite at nearly the speed of light.

As technology increases our reach, we are less and less required to move our bodies to receive or to dispatch information or goods and in general to further our civilization. Once, for instance, we had to leave the house to deliver a distant message. Now it is possible not only to phone, but to cook, grind coffee beans, answer the front door, sound the burglar alarm, turn off the lights, and view the world on television—in short, to run our domestic and public lives—without leaving the house. Once, mapmakers had to walk the land they charted. Now technicians sitting at a console obtain satellite photos of a continent simply by moving their fingers on a keyboard.

Once, microscopic spirochetes had to swim furiously for their lives. Now, millions of years later, packed into an organ called the brain, their nucleotide and protein remnants conceive and direct the actions of a highly complex amalgam of evolved bacterial associations called a human being. Perhaps groups of humans, sedentary and packed together in communities, cities, and webs of electromagnetic communication, are already beginning to form a network as far beyond thought as thought is from the concerted swimming of spirochetes. We stand no more chance of being aware of the totality of such a form of group organization than do the individual components of brain cells—microtubules, the putative remnants of spirochetes—understand their own mission in our human consciousness.

Although the spirochete hypothesis has not been proven, it suggests a symbiosis even older than those of both chloroplasts and mitochondria with the invaded host cells. We can date the symbiosis of spirochetes to before the time when prey cells and oxygen-breathing bacteria teamed up to make the first cells on their way to becoming nucleated. This is because today there are certain protists, such as *Trichonympha*, that survive from the time before respiring bacteria and after spirochetes had invaded host cells. *Trichonympha* has mitosis, which leads us to believe it has spirochete motility systems intact, yet it lacks mitochondria. Before the oxygen breathers caught on and began to develop a membrane around the nucleus, anaerobic spirochetes were already proliferating within our ancestral cells. They dragged the chromosomes along in a near-perfect system of reproductive distribution. And by maintaining the ability to produce an undulipodium, they preserved their option for rapid cellular swimming—an option that is exercised each time sperm cells rush out in their mad search for an egg.

CHAPTER 10

The Riddle of Sex

A sperm cell swims to meet an egg. Its tail, really an undulipodium, propels it along. Chemical changes at the surface of the egg cell allow the sperm to enter. Then, like the spent stage of a rocket ship, the undulipodium breaks off. The head, housing the nucleus of the cell, enters and ultimately fuses with the egg nucleus to produce a single new nucleated cell with twice as many chromosomes.

This is part of the cycle of meiotic sex, the typical sex of cells with nuclei. Whereas prokaryotic sex probably evolved in the Archean Eon more than 3,000 million years ago, meiotic sex evolved in the protists and their multicellular descendants, organisms with nucleated cells that are not fungi, animals, or plants. Meiotic sex probably evolved in the late Proterozoic Eon, about 1,000 million years ago. It was certainly present prior to the Ediacaran soft-bodied animals, who thrived approximately 700 million years ago.

At first—even second—glance, this kind of sex seems a superfluous and unnecessary bother. It has none of the virtues of the free bacterial genetic transfer associated with the world-wide microcosm. In the economic terms that biologists have used to describe it, the "cost" of this kind of sex—producing special sex cells with half the usual number of chromosomes, finding mates, and timing and performing the act of fertilization—seems all out of proportion to any possible advantage.

Sexually reproducing plants and animals are at the moment supremely successful groups—there is no doubt about it. And biologists have invented a just-so story to explain the existence of sex: that it is maintained because it enhances variety and speeds up evolution. But we believe that complex plants and animals have been preserved so far for reasons not directly related to biparental sex. Indeed, plants and animals have spread throughout the world in spite of two-parents sex.

Sex, in the biological sense, as we pointed out earlier, means simply the union of genetic material from more than one source to produce a new individual. It has nothing to do with copulation, nor is it intrinsically related to reproduction or to gender. According to this strict definition, the passing of nucleic acid into a cell from a virus, bacterium, or any other source is sex. The transfer of genetic particles such as viruses among different bacteria is sex. The fusion of two human germ cell nuclei is sex. Even the infection of humans by an influenza virus is a sexual act in that genetic material inserts itself in our cells. Symbiosis is like sex in that the genetic materials of different individuals eventually join in the formation of a new individual. Symbiosis, therefore, is considered "parasexual."

In the evolutionary lineage that led to us, two distinct happenings, originally having no causal relationship to each other, became interlocked in a feedback loop. One event was the reduction of the number of chromosomes in the nucleus of offspring cells; the other was cell and nuclear fusion. Both processes were originally accidents with no correlation to each other or to reproduction.

But these two processes were eventually connected, and since cellular fusion implies the mixing of DNA from sources separated in time and space, it is a form of sex. Finally, this particular sort of mixing of genetic material from precisely two sources became linked to reproductive mechanisms, arriving at the biparental sexual reproduction that many have held so crucial to the later evolutionary process. But we doubt that it has been so very crucial. It is far more likely that plants and animals themselves, with their elaborate tissue differentiation and highly complex forms, have been preserved by evolution. And complex bodies seem to be connected quite directly to meiosis, to DNA, RNA, and protein synthesis during the pairing of chromosomes in meiotic cell division, while only indirectly to the two-parent aspect of sex. We require reproduction through meiotic sex because our ancient, single-celled ancestors survived by way of meiotic sex. They got caught up in it to the extent that they could no longer reproduce asexually. Unable to swim and divide at the same time, our reproducing ancestor cells were encumbered by attached body cells condemned to evolutionary death. But since these were the same cells that gave them the advantages of fast swimming, large size, and effective food gathering, our ancestors became so involved in two-parent sex that soon they could not reproduce without it. From the beginning, the essence of animal existence involved two-parent sex. But two-parent sex itself was never maintained by natural selection. Indeed, if the evolutionary process can bypass biparental sex—through parthenogenesis in beetles, cloning in humans, or any other way—and still preserve complex multicellularity, it no doubt will. Biologically, sexual reproduction is still a waste of energy and time.

• • •

In human sex life (and in that of other plants and animals), two fundamental complementary steps are necessary for reproduction. First, the number of chromosomes in some cells must be reduced by exactly half—a process of cell division called meiosis. This creates either eggs or sperm. Second, these cells are then brought together and fused in the act of fertilization, thus restoring the normal number of chromosomes in the creation of a new cell, which divides to become a new individual. "Mating types" or what we call "genders" were actually a later refinement: the first germ cells pretty much looked identical, but over time developed into small moving sperm cells and large stationary eggs, which then had the added problem of recognizing each other in order to unite. Today all these conditions must be satisfied before a new sexually reproducing individual can grow.

But there is a riddle in all this. Why must two halves come together to make a whole only to become two halves again? We think that meiosis, the systematic creation of haploid cells (cells having half the number of chromosomes normal for each cell in an animal's baby), was prompted by that old menace of life on earth, starvation.

Faced with starvation, our ancestral protists would not have hesitated to resort to cannibalism. In some instances, the prey may not have been entirely digested, leaving the organism with a double set of genetic material, or in a "diploid" state, as cell biologists say. In other words, after eating a comrade cell, the two cells would fuse, and then so would their nuclei. Another way to create a diploid nucleus before the evolution of fertilization would have been if nuclei divided before the rest of the cell and then fused again when the rest of the cell did not follow in division. Even today many cells can't quite get mitosis straight. Many microscopists have seen cells make mistakes; they start to form two nuclei by division, then afterward the two nuclei fuse again. Accidental diploidy, whether by merging with potential food or part of oneself, must have happened all the time in the microcosm.

However it came about, diploidy was originally a monstrous, abnormal, swollen condition. To relieve themselves of this burden, affected cells had to devise means of reversing the process: bequeathing only half of their genetic material to their offspring when again they divided. This meant reworking the usual process of mitosis in which new cells are produced with the same number of chromosomes as their parents. It meant evolving the process of *meiosis,* where only half of the parent's genes, only one set of chromosomes, reached the next generation.

In modern meiosis, like chromosomes from each parental set line up together along the center of the spindle. If the cell were to divide at this point, each taking half the chromosomes, there would once again be two separate cells, each with one set of chromosomes. But before division, the chromosomes in each set replicate, leaving two sets of chromosomes lying alongside each other. Then the cell divides.

The kinetochores that button each pair of chromosomes together may still be marching to their own spirochetal drummer. They are not yet ready to divide themselves, so as the cell divides, the chromosomes going with each half are still paired. Finally, in each of the two offspring cells, the kinetochores divide, splitting the chromosome pairs and leading the cells to divide again, resulting in four cells. Each of these has but one set of chromosomes.

Experiments along these lines were probably repeated millions of times in different lines of protists, with most of the chromosomally diminished cells dying off, unable to produce all the proteins they needed to survive. Swollen cannibals relieving themselves of their doubleness could not also relieve themselves of any necessary genes or they would die. Those, however, that divided their genetic endowment precisely in half, returned to their original haploid condition of only one set of chromosomes. Under certain conditions, such as extreme dryness, chances of survival were greater in a doubled state. In times of active growth and searching for food, however, the ancestral haploid state was preferable. Alternating environments, such as the cyclical dryness of intertidal areas, favored organisms switching between haploid and diploid states. Eventually, some discovered that to remain viable, they had to keep switching between meiosis and cell fusion. A special ability to become more complex accompanied this periodic switching between haploid and diploid states.

So began the strange system of sexual reproduction. Even today, starvation, drought, darkness, nitrogen salt deficiency, and other privations induce cannibalistic fusions in many protists. On the face of it, dividing the number of chromosome sets only to double the number later seems a waste of time. But such time-wasters are often the only route to survival. They become ingrown, institutionalized, as in a person who learns a long way home and always takes it because he is unaware there is a quicker way to go. In the ancestral animal cells, the increased complexity reinforced such useless and annoying side trips. Among those cells cyclically switching between fused and original states were our ancestors.

Because many protists and nearly all animals, fungi, and plants have some sort of meiotic sex, it seems that meiosis and fertilization developed early in the history of nucleated cells. The protists that first experienced cannibalism and a return to haploidy already had their spirochetal mitosis system in place. Each offspring protist received a full dose of chromosomes each time its parent divided. Good mitosis in any lineage is a prerequisite for meiosis.

As Harvard Professor L. R. Cleveland, working alone in 1947, realized, a change in the timing of the replication of the chromosomal kinetochores was really all it took to evolve meiosis from mitosis. But Cleveland, with his Mississippi accent and idiosyncratic ways, never convinced his colleagues that he had solved the problem of the origin of meiotic sex.

Accidents changing the timing of events are very common in biology. Cleveland himself filmed reproduction into four instead of two cells, as well as mating by threes. He was well aware that deviant cell behavior was more normal than abnormal. Although Cleveland's films and papers and his scenario for the origin of sex are very convincing, at the time he presented them his colleagues were not ready to listen.

Cleveland recorded the cannibalistic fusion of protists called hypermastigotes. In some cases he even saw their nuclei fuse, though they almost always died afterward. He documented variations in the replication timetable of the kinetochores, the "buttons" that connect chromosomes to the microtubules of the mitotic spindle. Such mistiming also usually led to death. Cleveland saw the numbers of chromosomes in cells grow abnormally large. He recognized meiotic sex as a series of accidents that became twisted and bound together. Cannibalism, nuclear fusion, intracellular variations in the timing of replicating organelles, environmental cycles first favoring haploidy, then diploidy—all these things led to death. To survive was circuitous; it was to become sexual.

Like other evolutionary successes, sexually reproducing organisms probably began to proliferate with gathering momentum as they interacted with the environment and coevolved with each other. Very different species of sexual organisms today have obviously coevolved in great intricacy. To be fertilized, honeysuckle must be visited by hummingbirds and the platform flowers of century plants must receive bats flying at night. Milkweed flowers harbor the larvae of milkweed moths as they mature.

Most asexually reproducing organisms, which can make good copies of themselves in practically no time without bothering to arrange for a partner, are still best suited to many environments. But sexually reproducing organisms, perhaps in part because they tend to be multicellular, take longer to reproduce. This lets us and those like us occupy a niche in time that is quite distinct from nonmeiotic organisms such as bacteria. Individual development takes longer and may be better suited to certain cyclically changing conditions. Also, since complex animals undergo a complex program of development from their conception in the form of a fused egg and sperm cell, they have special aptitudes for identity and death. They save their potentially immortal genes in the divided form of eggs and sperm, but in each life these cells meet and reproduce to form a body of extra cells, an individual body which will die, often after fulfilling its unasked-for mission as a vehicle for the propagation of what are still essentially microbes. Mortality and identity do not depend so much on the complement of different genes we get from mom and pop, but on the process of meiosis itself. Meiosis may ensure that at least one copy of each important gene will be distributed to the next generation. Studies with ciliates show that a complex identity is not maintained by the influx of new genes but by meiosis: ciliate cells that fertilize themselves create an individual just as viable as one that gets its genes from a member of the opposite sex. Similarly, parthenogenetic animals do not need mates. But they must undergo meiosis.

People think of sex and reproduction as all of a piece. It helps restore perspective to remember that the three components of sexual reproduction that are now together once upon a time were separate. Reduction of chromosomes, fusion of nuclei from two sources, and the cyclical linkage of these processes in the sexual reproduction of complex individuals are actually at least three distinct processes. Each of these steps is essentially independent, and each has been observed in its own right. L. R. Cleveland recorded the nuclear fusion of two nucleated cells called hypermastigotes that had fertilized each other. Variations in the replication timetable of kinetochores, the dotlike attachment sites of chromosomes that may be another relic of spirochete motility systems, are also common. And many have reported species that have one set of chromosomes but resort to having two sets in certain environments. Meiosis or chromosome reduction and the complementary need for fertilization evolved in protists perhaps 1,200 million years ago. The connection of fertilization with embryo formation evolved later in animals, before some 700 million years ago. The strong linkage of reproduction to fertilization arose in mammals only 225 million years ago. So all the steps leading to our kind of sex—cannibalism and nuclear fusion, intracellular mistiming of various parts, and cyclical environments alternately favoring haploidy and diploidy—became tied together in the origin of multicellular eukaryotes.

Biologists have thought that sex persists because it increases the variety, the newness of offspring. This variation, it was reasoned, allows sexual organisms to adapt faster to changing environments than do asexually reproducing organisms. Yet there is absolutely no evidence that this is true. When the idea was tested by comparing animals that can reproduce either asexually or sexually, such as rotifers and asexually reproducing lizards, scientists found that as the environment varied, the asexual forms were as common as or even more common than their sexual counterparts. Females—rotifers or lizards—produce fertile eggs that grow into female adults capable of producing fertile eggs. No males can be seen, yet the populations of these rotifers and lizards are full of varying individuals well adapted to the exigencies of their environment.

Another theory holds that sex is important as a genetic rejuvenating mechanism. This theory is based on the observation that paramecia produced by asexual cell division survive for only a few months, whereas sexually conjugating strains survive indefinitely. But a certain paramecium itself refutes this theory outright.

Paramecium aurelia has one large macronucleus and two small micronuclei. The enormous macronucleus containing thousands of gene copies usually does all the work, making messenger RNA, while the diploid micronuclei do nothing. But when the paramecium gets ready for sexual conjugation but finds no partner, it undergoes a process called autogamy. Each diploid micronucleus divides twice meiotically, forming four haploid micronuclei. Then in nature's typically absurd style, all but one of the haploid micronuclei die. This last one divides mitotically again to create two micronuclei with exactly the same genes.

If a willing sexual partner is present after this getting-ready stage, conjugation occurs and one of the two micronuclei is sent to the partner in exchange for one of the partner's. The new and old haploid micronuclei fuse, completing the life cycle. But if no partner is available, the two genetically identical haploid nuclei mate with each other inside the same cell. They fuse. The paramecium ends up with one macronucleus and one micronucleus, but without having had biparental sex of any kind. No new genes have entered the paramecium in this self-fertilization. Yet the protist is genetically recharged and rejuvenated, able to reproduce again for generations.

All in all, our meiotic kind of sex is not the grand process we make it out to be. It is certainly far less important to the biosphere than bacterial sexuality, which is a strategy for immediate survival by which microorganisms receive new genetic components as easily as people catch a cold. Indeed, the life cycle of *Paramecium aurelia* shows that it is not the receiving of genes from a male and female—in other words, two-parent sex—that helps the paramecium survive. It is meiosis itself. We think that meiosis as a cell process was necessary for the evolution of complex animals. Even careful cloning of single plant cells produces highly variable offspring. With this in mind, it would seem doubtful that the meiotic sex cycle has been retained in complex organisms because it adds variety. Indeed, there are so many means of intracellular variation, (some probably owing to the former symbionts), that highly complex organisms need a way to make sure-they don't get swamped by it. Far from being a means of adding variation, meiosis is probably a stabilization mechanism.

Meiosis, especially the part that involves the pairing up of chromosomes side by side and special DNA, RNA, and protein synthesis, is, we believe, like a roll call or a taking of inventory. It ensures that all the sets of genes, including those of the mitochondria and plastids, are in order before the multicellular unfolding that is the development of the embryo begins. After all, animals and plants return every generation to a single nucleated cell.

Sex, like symbiosis, is one expression of a universal phenomenon, the mix-match principle. Two well-developed and adapted organisms or systems or objects combine, react, redevelop, redefine, readapt—and something new emerges. Human inventions continually exploit such mixing and matching. For example, simple wrist watches merge clocks and bracelets, tanks bring together trucks and cannon, synthesizers combine computers and pianos, and rotocrafts join the helicopter to the airplane. Whether expressed as a viral infection, the union of alga and fungus in a lichen, amoeba cannibalism, a marriage of two people, or the fusion of a video screen with a cassette recorder to make a videocassette, the principle of recombination is one that pervades life on earth. Yet it is reproduction, not recombination in the narrow sense of meiotic sex, that takes the upper hand.

Human beings are obsessed with sex because we feel pleasure when we indulge in it. But behind the scenes of sex lies the cellular imperative of reproduction. It is reproduction and not sexual intercourse per se which is really being spurred on by this form of positive feedback: repeating pleasurable experiences may lead to the mixing and matching of genes, but only because this side trip is presently necessary for mammalian reproduction. If "shortcuts" such as cloning are developed that bypass two-parent sex (and are accompanied by sufficient pleasurable feedback) complex animals would take them, reproducing faster while still exhibiting plenty of variation. In fact, cloned animals would probably show so much variation that they would need the fancy chromosome-pairing part of meiosis just to keep that variation under control.

CHAPTER 11

Late Bloomers: Animals and Plants

THE most primitive animal in existence, *Trichoplax,* was first discovered in 1965 crawling up the side of an aquarium in Philadelphia. *Trichoplax* is little more than a squashed pile of nucleated cells walking on microtubular 9 + 2 cell whips. No ancestors of *Trichoplax* have been identified. Because an egg is fertilized by a sperm and the fertile egg forms an embryonic ball of cells, we must call *Trichoplax* an animal. But that is about it: *Trichoplax* lacks head, tail, left and right sides. Superficially, the little creepy crawler is no more complex than a multicellular amoeba.

When a protist with cell whips first stuck to another cell and propelled it along so that the second cell's microtubules could be put to other uses, the evolutionary line that led to animals was launched. This was the start of cell specialization, which animals have raised to a high art. Some cells swam, others retained their capacity for mitosis and meiosis, still others committed their spirochetal apparatus to sensing the signs of the outside world. The antennae cells, balance organ cells, kidney cells, brain cells, mechanical receptors, and olfactory cells of animals never divide upon maturity. This is probably because the special uses to which they have committed their microtubules preclude the use of them as a mitotic spindle.

Accidents in which offspring cells do not separate but stick together or amalgamate are a common evolutionary occurrence. Multicellularity of one kind or another has appeared many times in many lineages of life, leading to larger, more complex organisms based on a recognizable cell plan. By no means are all the creatures of the microcosm single-celled. Myxobacteria, cyanobacteria, and actinobacteria are all primarily multicellular. Fungi are multicellular yeasts and many green seaweeds are multicellular *Chlamydomonas* algae. Animals and plants cannot be distinguished simply by their multicellularity.

Among the protists, the tendency for offspring cells to stick together led to the formation of multicellular amoebae such as slime molds; multicellular algal protists, including kelps and red seaweeds; and multicellular collared mastigotes such as sponges. Multicellularity probably evolved a dozen separate times in bacteria and more than fifty separate times in protists. Three kinds of protist organization were so spectacularly successful in numbers and variation from today's point of view that we raise them to the lofty status of kingdoms: the animals, the fungi, and the plants.

Usually the cells of a multicellular being are derived by cloning a single parent cell. But in some organisms (such as slime molds, gliding bacteria, and the ciliate *Sorogena*), cells from different sources but of the same species aggregate. Whether cloned or convened, the cells touch, subtly interact, and produce some orchestrated enterprise, such as a bacterial tree, a slime mold slug, or a ciliate spore tower.

All five kingdoms—bacteria, protists, fungi, plants, and animals—have multicellular members. But in the first four kingdoms—even among plants—the multicellular organisms have only minimal strands of communication between their body cells. In the kingdom Animalia, on the other hand, multicellularity and interactions between cells became an expertise. Animal multicellularity is exquisitely refined and organized. The animal cell is highly specialized and bound to its neighbors by a variety of elegant intercell connections—septate junctions, gap junctions, tight junctions, and desmosomes, to name a few. It is these cell junctions, whose distinctions have only recently become visible via the electron microscope, that determine the extent and quality of communication between cells. Along with the blastula, the ball of cells that becomes an embryo, esoteric cell junctions are the true marks of animality.

Early in their development during the Proterozoic Eon, about 1,000 million years ago, their appearance must have distinguished the very first animals from the multitude of sticky multicellular microbial colonies. From intense evolutionary interaction in closely knit communities came groups of cells with such mutual coordination and control that we can hardly imagine that they came from mixed populations of microbial strangers.

The first algal ancestors to plants were little more than chains of cells full of chloroplasts that lived in a thready mass. Plant spores are known to have been ashore by about 460 million years ago. (Compare this to the more ancient evolution of animals, which occurred about 700 million years ago, though animals came ashore only 425 million years ago.) We can trace probable steps in the evolution of plantlike cells from algal ancestors by looking at the lives of volvocines, a type of green pond weed. Each volvocine cell has two 9 + 2 whips, an eyespot, and a chloroplast. The single cells of the colony are suggestively similar to the protist *Chlamydomonas*, a unicellular alga. Four volvocine cells arranged together in a stable gelatinous disk are recognized as the organism *Gonium sociale.* Yet each cell in *Gonium sociale* is capable of swimming off on its own to start a new colony.

More complex than *Gonium* are intermediate species, organized forms of colonial algae consisting of 16 or 32 volvocine cells. The most highly organized is *Volvox,* a hollow sphere composed of half a million volvocine cells. *Volvox* rotates on its axis as it swims and, unlike the less organized volvocines, only a few specialized cells near the rear pole are able to divide to produce offspring colonies. Offspring cells break off from their single parent and move inside its hollow interior, where they continue to divide, forming miniature multicellular spheres of their own. Eventually the offspring spheres release an enzyme that dissolves the gelatin holding the parent sphere together. The parent sphere breaks apart and the new colonies pop out. Although this is reproduction without sex, some *Volvox* species turn sexual. Elaborate green translucent spheres release eggs or sperm; some hermaphrodite spheres release both at the same time. The released cells, which resemble the original green algal ancestors, find each other and fuse, starting with their cell whips. After fertilization is complete, the cell whips are withdrawn back into the fused cell and meiosis begins, followed by mitosis. The results of these processes are many offspring cells which stick together and grow into new *Volvox* colonies.

Algae dwelled in wet, sunlit shallows. Occasionally these shallows dried up, and those algae that could remain wet on the inside while dry on the outside had the evolutionary edge. They survived and multiplied to become the early plants—low-lying forms without stems or leaves, related to modern-day mosses and liverworts, that could not support their own weight out of water. Algae became land plants by bringing water with them. A watery medium was required by these earliest plants for their two-tailed sperm to swim to their eggs. The development of fertilized eggs into embryos inside the protective covering of parental tissue distinguished them as true plants, rather than protist algae colonies.

Dry land was as hostile an environment for plants as the moon is for us. It was crucial for the gelatinous tissues of algae not to collapse or dry out. Becoming terrestrial meant developing a tough, three-dimensional structure as contrasted to earlier low-lying forms. Using atmospheric oxygen (that old waste product of the cyanobacteria), the early plants evolved a cell wall material called lignin. It is this lignin that, combined with cellulose, gives such strength and flexibility to shrubs and trees. This sturdiness led to the development of a so-called vascular system transporting water up from the roots and food down from the flattened ends of branches that were early leaves. The new vascular plants also differed from their predecessors by having two sets of chromosomes in their cells instead of one.

When the microcosm bloomed into the macrocosm in the form of beautiful plants, the invisible microbes were everywhere. The resistance to drying out, the production of lignin, and the conquest of land probably involved microbial symbioses.

The Canadian biologists K. Pirozynski and D. Malloch believe plants emerged from symbioses between algae and fungi. According to these scientists, they were like lichens in reverse: in the symbioses leading to plants it was the algal, rather than the fungal partner that was dominant. It is probably no coincidence that 95 percent of today's land plants have symbiotic fungi in their roots, called mycorrhiza. Many plants will shrivel and die if deprived of their fungal partners in the soil. (Most of the 5 percent of plants that lack persistent fungal symbionts have taken again to the water; they are aquatic plants.) Indeed, the difficult conquering of dry land is easier to understand if it was done by symbionts, cooperative partners that collaborated to do what neither could do alone. And the discontinuous evolution, from soft, wet algae to lignified land plants, again suggests symbiosis.

By 400 million years ago, as the first jawed fishes ambled ashore and at the time of the earliest wingless insects, vascular plants were already thriving. A crucial challenge of the new land environment was water shortage. One solution was the development of seeds. The durable seed permits the plant embryo to wait for the best moment to develop. The invention can be appreciated by conjuring up a comparable structure for mammals. Imagine if human zygotes were wrapped in protective time-release capsules that were activated only by a booming peacetime economy. How convenient it would be for a distracted young woman if she could collect and store her babies-to-be, germinating them only after receiving her college diploma, buying her house and, in general, securing her future. Seeds permitted embryonic plants to wait and silently monitor the environment until favorable conditions arose. Fertilization and development within the moist tissues of the parent enabled seed plants to thrive despite the inconstancy of the rains.

The first forests contained giant "seed ferns." These were trees that looked like overgrown ferns but which, unlike ferns, produced seeds. For over a hundred million years, from 345 to 225 million years ago, during the time of the evolution of winged insects, torpedo-shaped squids, and dinosaurs (among others), the seed ferns grew in lush, tropical splendor. Forests of great trees, neither truly ferns nor flowering plants but cycadofilicales, a group unto itself, the seed ferns stretched out across the vast expanses of land. They swayed in warm winds from the southern reaches of Gondwanaland to the rolling hills of Laurasia. Occasionally crushed underground by moving continents, they also formed the material basis for the world's richest coal fields.

About 225 million years ago, the seed ferns were supplanted by their descendants, the conifers or cone-bearing plants which were the main diet of some of the first vegetarian dinosaurs. The cone-bearing plants, like the modern pine and spruce, probably were hardier than seed fern trees because of their greater tolerance to cold. Glaciers appearing on several continents 225 million years ago are thought to have presented conifers with adaptive opportunities. The seed ferns and other plants that left their imprints in coal were largely tropical. The naked seeds of the seed ferns were not enclosed in any kind of fruit and so the seed ferns must have been very sensitive to the cold. They did not last the long winters of glacial times. For many species of conifers, however, such as today's evergreens, subzero temperatures did not present a major problem. With a fungal microcosmic network at their roots—sometimes popping up as mushrooms—the ancient evergreens, too, expanded into the forbidden regions of alpine heights and blizzarding extreme latitudes.

The successors to the cold-tolerant conifers were plants with flowers. They were descendants of the same early land plants that also produced the seed ferns. Judging from certain palmlike fossils, they had already evolved 123 million years ago and shortly after, by 114 million years ago, had rapidly spread to all parts of the world's land. The most successful manifestation of chloroplasts on Earth, flowering plants coevolved with animals from the beginning. The suitor's bouquet is only a recent manifestation of this age-old intimacy.

Insects multiplied on sweet nectar and took other floral kick-backs in return for the service of cross-pollination. Birds and mammals stopped some of the rampant growth by eating the fruits and leaves of flowering plants, but the flowering plants or angiosperms fought back by evolving molecular toxins and hallucinogens to ward off the animals, as well as developing hard fruits (nuts) and seeds (pits) in order to protect their embryos from animal digestion. A side effect of these protective countermeasures was an effective animal distribution system for enclosed flowering plant embryos.

It may not be a coincidence that the first mammals, warm-blooded egg-layers and small marsupials, date almost exactly to the period of the first flowers, evolving some 125 million years ago. Quick-thinking, vegetable-eating mammals provide another striking example of cooperation in evolution, as their interest in plant foods probably led them to become well-fed disseminators of angiosperm seeds. Some flowering plants such as banana and orange trees no longer make functional seeds and have the most remarkable dispersal strategy of all: season after season human growers clone them. The plants are asexually perpetuated because they taste good. Our seniors on land, plants indeed seem very adept at seducing us animals, having tricked us into doing for them one of the few things we can do that they cannot: move.

The structure of plants seems to preclude evolution toward complex behavior. Yet as we gaze back on their impressive record, we may begin to suspect plants of being a bit less helpless than they seem. Our central nervous system and brain evolved as an adaptation to the eating of plants, and the eating of the eaters of plants. Plants don't really need brains; they borrow ours. They have a strategic intelligence that resides more in the chemistry of photosynthesis and the ploys of the genes than in the tactics of the cerebral cortex; we behave for them. And what is behind this different wisdom of plants? It is no more nor less than the ancient microcosm. Microbes, in the form of chloroplasts, mitochondria, and the agitated residue of spirochete motility systems, provide the basis for botanical success.

Animals approached the emergence from water to land a bit differently. Perhaps because of their poorer heritage—the presence in them of only mitochondria and spirochete systems but not chloroplasts—it took animals 35 million years longer to come ashore, even though they had evolved before plants.

Lacking hard parts, the very first animals are preserved only in a limited fashion, as whole body impressions in sandstones; trapped long ago on sandy beaches, they were buried before they could be eaten by enterprising bacteria. These first animals hatched progressively larger globular and wormlike forms with even more cells. Cells stayed together even longer after dividing. New ploys for survival were exploited. Some, for instance, used their 9 + 2 cell whips as cilia to sweep food particles, bacteria, and small protists, into their digestive tracts. With time, bigger and better food-collecting animals evolved, all developing from blastulas. Some of them were probably important architects of the modern living landscape. In human terms these animals were still not large. Yet communities of them together built enormous coral reefs, while others dined on the soft microbial mats that were widespread during the Age of Bacteria. Today microbial mats exist only in isolated and remote parts of the world not favored by eukaryotes, such as the extremely salty environments of intertidal zones. (These ocean edges, unmolested as they are by browsing snails, disruptive worms, or condominium developers, provide important sanctuaries for the world's oldest residents.)

The first sandstone imprints, left by the oldest animals we know about directly, were found in Ediacara, near Sydney in Southeastern Australia. Similar 700 million-year-old soft-bodied animal fossils from all around the world are also known as *Ediacaran*. Some very successful animal phyla are already present in the Ediacaran remains. The major phyla which evolved from the early Ediacaran animals and their relatives around the southern hemisphere were the soft, segmented sea worms called annelids. Whether or not remains of the extremely successful arthropods (ancestors of many segmented animals with external skeletons, including trilobites, crabs, shrimps, cockroaches, and lobsters) and the coelenterates, (which stung their prey with poisoned tentacles bearing overgrown undulipodia nearly a millimeter long) are present in Ediacara is debatable. Certainly many animals now extinct were then present.

Other important groups of skeleton-less animals, such as the echinoderms (the phylum to which starfish and sea urchins belong) may also have existed by 700 million years ago. But lacking hard parts, they are difficult to recognize for sure until 560 million years ago. Nonetheless, we are fortunate that many soft-bodied forms have been nicely preserved amidst the violence and tumult of geologic history. The Ediacaran fauna, first brought to the attention of the scientific world by the late geologist Martin Glaessner at Sydney University in Australia, are a thrilling find, similar to finding ruins in a wind-swept desert or stumbling upon ancient Khmer temples in the Cambodian jungle. We now understand much more clearly the great "delay" in the fossil appearance of animal life. Soft animals of great complexity preceded animals with hard parts, just as soft people making artefacts of great complexity preceded knights in armor with their hard parts. The very first animals—little societies of cells, spongy beasts, and early *Trichoplax*-like forms far less complicated than the Ediacaran animals—must never have fossilized at all. The descendants of early animal communities such as *Trichoplax* are so watery, delicate, rare, and generally inconspicuous that they defy study even today. Without hard parts or internal skeletons the earliest animals, like most protists, completely disintegrated at death. Their structured descendants, however, turned chalk and silica to gorgeous shells and skeletons, leaving to posterity a wealth of clues about early life in this, our Phanerozoic Eon.

The opening of the Cambrian Period, some 580 million years ago, represents the coming of the modern age, the Phanerozoic Eon in which we live. During the Cambrian, fossils of such clarity and abundance appear all over the world that for a long time life itself was thought to have begun only then. Until the fossil microcosm was revealed during the last two decades, older rocks were lumped into an enigmatic and enormous period of time called the Precambrian, which extended from the beginning of the Earth. Today we know that life did not begin with trilobites and similar well-developed animals of the Phanerozoic. What began in the modern Phanerozoic Eon were skeletons and the development of shells, rigid coats, and other hard parts to ward off waves and predators—all of which had the side effect of leaving a clear record of animals in the fossil record.

Many seventeenth- and eighteenth-century European naturalists called certain intricately patterned fossils "figured stones." Some considered them evidence of surfacing animal "forms" that were continuously produced underground, while others suggested they dated from Noah's Flood. John Woodward, for instance, Professor of Physic at Gresham College in London, thought that on the basis of English plant fossils he could estimate just when during the year it was that the Great Flood occurred. As he wrote in 1695 ". . . of all the various Leaves, which I have yet seen, thus lodged in Stone, I have observed none in any other State, nor Fruits further advanced in Growth, and towards maturity, than they were wont to be at the latter End of the Spring season." "The Pine Cones," he went on to say in another work, "are in their vernal State; as are all the Vegetables, and the young shells. The Deluge came on, and a stop was put to their further growth, at the end of May." Well into the nineteenth century, fossils, indeed all of nature, were interpreted in a biblical context and studied with an eye to the verification of the word of God. But the French naturalist Georges Louis Leclerc de Buffon, the British Erasmus Darwin (Charles's grandfather), and later the French evolutionist Jean-Baptiste Pierre Antoine de Monet de Lamarck, the English author and artist Samuel Butler, and others had already begun thinking about these rocks in a different way. The London encyclopedia publisher Robert Chambers anonymously advanced the idea in his 1844 *Vestiges of the Natural History of Creation* that God did not separately create species and actively oversee them, but that He had created life on Earth only once, in the beginning, and had let it run its vital course.

New ideas about life spilled over into changed ideas of God. Charles Darwin, a man who at first wrote he did not "in the least doubt the strict and literal truth of the Bible" was later to wonder about "Ichneumids tearing their prey to bits, how could a beneficent God have willed it so?" In 1844 Darwin wrote his colleague, the botanist Joseph Hooker: "At last gleams of light have come, and I am almost convinced (quite contrary to the opinion I started with) that species are not (it is like confessing a murder) immutable."[43] By the time the theory of biological evolution became widely known in the late nineteenth century, Cambrian fossils were sometimes reinterpreted as representing God's first, general creation of all so-called lower life forms. With the discovery of bacterial, protist, and Ediacaran fossils, however, it has become clear that life's "sudden appearance" in Cambrian rocks was an illusion that had to do with animals forming hard outer jackets at this time. Our ancestors had been around for about 3,000 million years before they started forming shells and external skeletons visible to the naked eye.

Animals' first hard parts, dating to some 580 million years ago, were formed of calcium phosphate and the organic material called chitin. Chitin was used by both trilobites and another line of rampant Cambrian arthropods, the eurypterids. Sometimes reaching over three meters in length, all of these giant "sea scorpions" are now completely gone. Other Cambrian forms included the brachiopods, clamlike creatures known as "lamp shells." Nearly all brachiopods are extinct, but some of these ancient mollusks still may be seen by divers along the Atlantic seaboard of North America.

Perhaps the most famous of the Cambrian fossils are those of the Burgess Shale found by the geologist Charles Walcott near Field, British Columbia in 1910. Walcott assembled a team to dynamite out the area and exposed a meter or so of extremely fossiliferous shale which had been the muddy flats of a long-gone sea. At present the former flats are high on a mountain pass in the Cascades, a mountain chain extending from the states of Washington and Oregon into western Canada. Walcott and his colleagues took some 35,000 fossil specimens over the next decade, many of which are now at the Smithsonian Institution in Washington, D.C.

Nearly all of the Cambrian fossils belong to one of the thirty or more well-known phyla into which present-day animals are categorized. A few enigmas (often with the scientific label "Problematica") are left in the paleontological literature and described without classification.

A very widespread group of fossils found in great numbers

from lower through middle Cambrian rocks are the archeocyathids. They are cone-shaped skeletal structures. Extensive communities of them dominated the tropical seas. These rocky, coned beings are unknown in any form whatsoever today, and speculation, as it is wont to do in a field lacking information, runs wild. Were they corallike, or spongelike, or were they not even animals at all? Whatever they were, the archeocyathids abruptly disappeared in the upper, or more recent, Cambrian. All of their numerous families, species, and genera became extinct fewer than 80 million years after they had first appeared. Their fossils are known from locations in England, Siberia, Canada, and elsewhere.

Other examples of biological extravagance, of ghost species dimly recorded by an incomplete fossil record, are also known. "Things" that look like tiny engraved designs in limestones are called stromatoporoids. Beginning in the late Cambrian Period and persisting into much of the rest of the Paleozoic Era, at least one paleontologist has claimed these were vast fields of blue-green bacteria. Others believe stromatoporoids were communities of sponges or spongelike animals. So no one even knows whether these problematic stromatoporoid creatures were bacteria, or animals, or neither.

Most people if they strolled along Cambrian beaches, and looked at the ocean rather than at the low-lying, land-based microbial mats, would not really notice much difference between the half a billion-year-old fauna and those of today. The minor differences between forms of marine life are a matter of taxonomic esoterica anyway. What is more, certain organisms, such as *Limulus*—the horseshoe crab—have not changed much at all since Cambrian times. Horseshoe crabs are not crabs but another kind of arthropod altogether. At Cape Hatteras and Cape Cod *Limulus* lives on, its progeny scattered along the Atlantic coastline. Stuck firmly together in mating pairs, slowly ambling toward the water, or exposed on its back with legs flailing, the horseshoe crab looks just as it did 400 million years ago.

To the studious eye and receptive mind, however, Cambrian bathing would no doubt offer its own peculiar delights; scuba diving might be even more thrilling. Every species found, if time could be turned back, would be different in detail from species today. Not a single Cambrian species—and thousands are known—is still alive today. All the familiar modern forms, such as the blue mussels, lobsters, oysters, quahogs, and crabs, would be completely absent. There would not be a single fish. Instead, waving worm communities, witch-hat archeocyathids, mysterious stromatoporoids, bunches of brachiopods, solitary corals, and obscure sponges would abound. Whole phyla of giant, quivering, globular marine beings may have multiplied in copious abundance. Formed entirely of soft gelatinous parts, such creatures would have left no good fossils. Emerging from the water, our Cambrian diver would witness the difficult struggles of that minority of threatened animals caught panting, drying out on shore—the ancestors of those who would adapt to land.

In hindsight, the adaptation of animals to life on land can be seen as an engineering problem of daunting proportions, comparable in complexity to human beings living on other planets. Life on land had its attractions: there was much oxygen to be taken from the air and the land itself provided a frontier for expansion. But there were also extraordinary obstacles.

Emigration onto land and continued reproduction there required major changes in all of the organ systems of the animals that achieved it. Withdrawn from the buoyant medium of water, organisms collapse under their own weight. So stronger muscles and sturdier arrangements of bones were needed to compensate for the absence of buoyancy. Breathing equipment functioning in gaseous air was also imperative: oxygen that in water was present in a few parts per million now, in the atmosphere, jumped in concentration level several thousand times. In addition, surface coatings such as skins, cuticles, and carapaces were necessary for animals to withstand the harsh, unfiltered sunlight that shone directly on land. But the greatest threat of all was drought. On land there persistently lurks the deadly threat of desiccation, of death by drying out. This choking thirst had to be overcome. Organisms that could not somehow carry water with them as they emerged from water were doomed.

The move to terra firma was probably prompted by geological upheavals, by the periodic receding of waters in which animals lived. Indeed only a very few representatives of the animal kingdom have ever successfully adapted to land. In not a single phylum of animals are all members land-dwelling. Animals that today can complete their life cycle on dry land may be traced back to three or four groups of ancestral aquatic animals that came ashore: some arthropods, namely the insects and spiders, a very few mollusks, such as the land snails, some annelid worms, and some members of our phylum, the chordates. Chordates have a tube-shaped nervous system toward the back of their bodies and always develop gill slits at some time during their life cycle—evidence of an oceanic ancestry. The subphylum of chordates to which we belong is the vertebrates. Of the other thirty or so invertebrate groups, not a single member of any species has ever survived through its normal life cycle confined to dry land.

Of course many aquatic animals have piggy-backed their way onto land. Living in the moist nasal cavities and guts of vertebrates, pentasomes and tapeworms, for instance, have only superficially colonized that one-third of the earth which is really earth as opposed to ocean. Even in the six classes of chordates—bony and cartilaginous fishes, amphibians, reptiles, birds, and mammals—only the three most recently evolved—reptiles, birds, and mammals—have ever completely eliminated the necessity to be submerged in water at some time during their life.

And among those species alive today whose ancestors were ocean-dwellers with hard parts, there are those that have reverted back to a maritime environment. The swimming ichthyosaur reptiles of the Mesozoic Era evolved on land and only later, under a different set of selection pressures, returned to the ocean. So, too, the seals and sea lions we admire at the aquarium are in truth no more than "water dogs." A far evolutionary cry from fish, the ancestors of seals, sea lions, dolphins, and whales all evolved as four-legged land creatures. We land organisms all derive ultimately from aquatic ancestors. Fertilization betrays a common aquatic ancestry for every living animal. The essential act of animal creation still always takes place in water. Derived from sea, river, pond, or the body's own tissue fluid, sperm and egg always meet in a wet environment.

In a sense, those animal species that fully adapted to the land did so through the trick of taking their former environment with them. No animal has ever really completely left the watery microcosm. The blastula and embryo still develop in the primeval wetness and buoyancy of the womb. Waterproof shells and coatings encapsulate the primordial environment. The concentrations of salts in both seawater and blood are, for all practical purposes, identical. The proportions of sodium, potassium, and chloride in our tissues are intriguingly similar to those of the worldwide ocean. These salts are compounds which animals took with them as they made their perilous voyage onto land. No matter how high and dry the mountain top, no matter how secluded and modern the retreat, we sweat and cry what is basically seawater.

Crucial to their transfer onto land was what animals did with the element calcium. Calcium is a raw material in the making of many of the most magnificent biological structures, such as the human skull or the White Cliffs of Dover. The amount of calcium in solution in the cytoplasm of a nucleated cell must always be kept around one part in ten million. Yet calcium in seawater can be 10,000 times or more higher than this. Calcium tends to rush into cells, causing them to be continually ridding themselves of it. As all cells with nuclei do now, the first animal cells had to continuously export calcium outside their cells in order to stay healthy. Today, calcium carbonate is made by special cells inside membraneous sacs. The chalky substance is transferred in precrystaline form via channels—along which run the ubiquitous microtubules—to the outside of the cell.

Calcium plays a central part in the metabolism of all nucleated cells. It plays an indispensable role in amoeboid cell movement, cell secretion, microtubule formation, and cell adhesion. Dissolved calcium must be continually removed from the surrounding solution for microtubules to function in mitosis, meiotic sex, and brain activity. Because the "chemo-" part of chemoelectric messages sent by the nerve cells in the brain has largely to do with calcium, the neuron-firing communication networks of the brain depend as much on calcium as telephone communication does on copper telephone wire. By 620 million years ago the first tiny animal brains had evolved.

Perhaps more important for these early animals was the use of calcium in the operation of muscles. Muscles contract when dissolved calcium and ATP are released in precise quantities around them. The calcium must be scrupulously kept in quantities far lower than those of seawater or chemistry takes over and the calcium phosphate comes out of solution in a solid form. (This is why athletes overworking their muscles tend to develop calcium deposits.) Muscle tissue, and the actinomyosin proteins comprising it, tends to be the same in all animals. The origin of actin is an evolutionary mystery; an actinlike protein has been reported in the putative ancestor to our cells, *Thermoplasma*. If confirmed we have still another case of an invention that originated in the bacterial microcosm.

The soft-bodied underwater worms and blobs of Ediacaran times swam using muscles. To do so they controlled their calcium metabolism. Since muscle contraction responds to calcium release, it is extremely probable that the early Cambrian sea creatures, even the earliest squiggling annelid worms, must surely have had muscles under calcium control. Like Greek and Roman breastplates and helmets, some of these early animals must have secreted bits and pieces of calcareous armor and protective films that were not yet full skeletons.

It is rather remarkable that in otherwise very closely related species, one will calcify while the other will not. For instance, the only difference between certain very closely related species of coralline red algae is that one is covered by stony calcium carbonate plates while the other is totally soft. Stephen Weiner of the Weizmann Research Institute in Israel believes that the calcifying species makes enough of the proteins having a regular spacing to fit the calcium carbonate crystals in the proteins' framework. The other species, however, makes too little or an altered form of the protein with inappropriate spacing. On the other hand, since in some cases separate species of organisms which branched apart millions of years ago will both produce calcium carbonate today, it is probable that the ability to precipitate calcium compounds in a regular manner has successfully evolved many times in many different species for many distinct purposes.

Always used by nucleated cells, excess calcium must be excreted or harmlessly stockpiled out of solution. Since Cambrian times organisms have been stockpiling their reserves as calcium phosphate, which takes such forms as teeth and bones, or as calcium carbonate, as in chalky shells.

Skeletons did not appear out of nowhere during the Cambrian: Ediacaran muscles preceded Cambrian skeletons. The need to continuously respond to calcium surpluses in the cell made it easy for some animals to stockpile calcium salts inside or outside their bodies in dump heaps that eventually became skeletons and body armor. Just as termite nests are largely constructed of insect excrement and saliva, so skeletons and teeth are made of compounds that originally had to be excreted as waste.

Most animal shells and outer coats today are composed of calcium carbonate. Tiny ocean protists such as foraminifera and coccolithophorids extruded so much calcium into the water over such long periods of time that they made that famous piece of English real estate, the White Cliffs of Dover, a towering deposit of limestone and chalk. (Like coal or oil, such organic carbon reserves are not wasted but held in biospheric storage until life discovers new ways of recyling them.)

The new organs that supplanted the old, waterlogged ones were forced to be successful. Gills, expert at culling oxygen from water, were useless in the air. Over the millennia they became relics, like the gill slits that look like tiny scars under the ears of human fetuses. Lungs which could deliver air to the circulatory system evolved in some chordates, such as the amphibians, reptiles, and mammals. An analogous system of air channels called trachea evolved in land-adapted arthropods such as spiders and insects.

When facing frightening environmental perils, organisms warded off the need for radical change by incorporating the new into the tried-and-true old. The assembly of bones that had evolved in swimming fish served later to support amphibians on land, and to aerodynamically support birds in the air. Calcium waste near muscles became basic construction materials. Early vertebrates evolved into fish—bilaterally symmetrical beings that were essentially escape artists and speedsters, darting away from predators and rapidly pursuing their prey.

Competition among vicious predators along with desiccation in shallow waters forced early animals to live on land. But the scorching Earth was no happy alternative to the warring seas. The land was in some ways an Edenic paradise, a sanctuary originally free of dangerous predators. But it was also a separate hell—an environment of torturous sun, biting wind, and decreased buoyancy. Calcified structures such as snail shells began as dumps for excess calcium but wound up as a combination of gravitational support structures, shields against sunlight and predators, and organic "aquariums" protecting against the dangers of desiccation.

In all that has been written about the evolution and expansion to land of plants and animals, it is easy to forget the role of fungi. Fungi represent a third fundamental organizational plan of evolving nucleated cells. Fungi develop from spores and grow slender tubes called hyphae, which are often divided by cross walls called septa. Part of their alternative organization includes the fact that fungal cells may contain many nuclei per cell, and that cytoplasm can flow more or less freely from cell to cell through the septa. Unlike animals with stomachs, fungi digest outside their bodies. They grow by absorbing nutrients as chemicals, rather than by photosynthesis or eating. Molds, morels, truffles, yeasts, and mushrooms are common examples of fungi. Of the estimated 100,000 species of fungi, most are terrestrial.

Though a separate kingdom, fungi are intimate with plants. They probably evolved from a line of funguslike protists that absorbed food directly from the living or dead bodies of algae, plants, and animals; and fungi seem to have coevolved with plants in the move to the land. Fungal fossils, found especially in fossils of plant tissue, are known from over 300 million years ago. Symbiotically associated with plant roots, fungi transfer the nutrients phosphorus and nitrogen on a worldwide scale. Without fungi, plants would starve for these mineral nutrients. For this reason it has been suggested that primeval forests would have been impossible without them. Resilient organisms, some fungi are able to grow in acid, while others can grow in extremely nitrogen-poor environments. Their chitinous cell walls, hard and stiff and able to resist drying out, made them well suited to life on land.

All fungi form spores. If desiccation threatens and partners are not available, asexual spores will form immediately without the bother of sex. On the other hand, many fungi indulge heavily in meiotic sex. Some can have up to 78,000 different mating types or "sexes" in a single species, as does the roadside mushroom, *Schizophylum commune*.

Fungi are indeed an underrated kingdom, in the past unnecessarily lumped together with bacteria and other nonanimals in the kingdom Plantae. But organisms like toadstools and yeasts exhibit such peculiar characteristics that biologists now agree they merit separate taxonomic status as the kingdom Fungi.

Fungi figure in human culture as well. Pigs in southern France are trained to sniff out truffles, which, considered a culinary delight, are so intimately involved with certain trees that no one can farm them. We ingest penicillin, a chemical defense against bacteria first made by the green bread and fruit mold *Penicillium* long ago. The drug penicillin prevents infectious bacteria from making cell walls. Thus this fungus, by saving itself, has also saved millions of human lives.

Yet fungi, like bacteria, are not always so benign. The ergot fungus can infect whole fields of rye, causing spontaneous abortions in cattle which eat the infected grain, and an incurable affliction known as St. Anthony's fire. In the Middle Ages, people who ate bread made from such infected grain were poisoned. Furthermore, ergot contains lysergic acid, the source from which the Swiss chemist Albert Hofmann first synthesized lysergic acid diethylamide or LSD. While experimenting with lysergic acid, a drug which is used in small amounts to induce childbirth, stop afterbirth, and aid coronary seizures, Hofmann accidentally discovered LSD, the most powerful hallucinogen known.

The specific fungal and fungi-related tendencies to interfere with disparate animal and bacterial metabolic processes suggest ancient evolutionary defense mechanisms. The psychopharmacological changes induced by hallucinogens such as those from the *Amanita* or the *Psilocybes* mushroom represent coevolutionary survival patterns of our animal predecessors. Imagine the intense selection pressure on our hungry mammalian ancestors. Early on, this pressure may have led to those who could metabolically process formerly toxic foods. Some of these foods were fungi and plants or fungi-plant combinations which were environmentally plentiful. The plants and fungi, in turn, had pressure on them to redevelop deterrent alkaloids like ergot. The animals that processed the harmful alkaloid-ridden food barely survived the metabolic experience. Mind- and body-altering agents may recall a sort of metabolic war—the imperfect coevolution of eater and eaten species.

While all this is speculative, it is clear that all species which have ever evolved have coevolved. This applies to the macrocosm, of course, no less than the microcosm. Although the development by plants and fungi of alkaloids to regulate their mutual growth rates or to deter animal predators can be looked at as a sort of chemical arms race, it also is reminiscent of the symbioses of the microcosm. Fungi are the major causes of disease in plants. But they are also crucial to plant survival. Vicious relationships between predator and prey can often be seen as part of larger symbioses in which even killing plagues aid the afflicted populations by not letting them overgrow their food supply.

There is really not so much that is new in our modern, Phanerozoic Eon (570 million years ago to the present). Except for such esoteric innovations as snake venoms, plant and fungal hallucinogenic poisons, and cerebral cortices, by the end of the Proterozoic Eon (2,500–580 million years ago) the evolution of almost every major survival technique which life uses today had taken place. Grass-green bacteria, spirochetal remnants, and mitochondria had long since set the stage. On a changing Earth, they preserved the ancient breakthroughs of early life by evolving into new forms even as they retained their exquisite abilities to make food from the sun, deploy genetic information, and combust dangerous oxygen. Some of the microbial recombinations were seaside animals able to control their calcium. But the actual shapes assumed by animals are almost incidental to the microbes. The important point is that by retaining their primordial environments in stable, water-resistant structures such as external skeletons, the microbes could, and did, expand. In doing so they left animal remains that, for a long time, were considered to be fossils of the first life forms on earth. Now we know better.

Life on Earth is such a good story you cannot afford to

miss the beginning. Do historians begin their study of civilization with the founding of Los Angeles? This is what studying natural history is like if we ignore the microcosm. Plants, fungi, and animals emerged from the microcosm. Beneath our superficial differences we are all of us walking communities of bacteria. The world shimmers, a pointillist landscape made of tiny living beings. Giant redwoods and whales, mosquitoes and mushrooms are intricate symbiotic networks, modular manifestations of the nucleated cell. By such proxies microbes found their way onto dry land.

CHAPTER 12

Egocentric Man

MAN is the consummate egotist. Before Copernicus founded modern astronomy our ancestors believed that their home, the Earth, was at the center of all the universe. Despite Darwin's demonstration that we are only one recent branch on an evolutionary tree, most people still believe that human beings are biologically superior to all other life. Mark Twain in his once-censored essay "The Damned Human Race" puts it this way:

> From this time onward for nearly another thirty million years the preparation moved briskly. From the pterodactyl was developed the bird; from the bird the kangaroo, from the kangaroo the other marsupials; from these the mastodon, the megatherium, the giant sloth, the Irish elk, and all that crowd that you make useful and instructive fossils out of—then came the first great Ice Sheet, and they all retreated before it and crossed over the bridge at Bering Strait and wandered around over Europe and Asia and died. All except a few, to carry on the preparation with. Six Glacial Periods

with two million years between Periods chased these poor orphans up and down and about the earth, from weather to weather—from tropic swelter at the poles to Arctic frost at the equator and back again and to and fro, they never knowing what kind of weather was going to turn up next; and if ever they settled down anywhere the whole continent suddenly sank under them without the least notice and they had to trade places with the fishes and scramble off to where the seas had been, and scarcely a dry rag on them; and when there was nothing else doing a volcano would let go and fire them out from wherever they had located. They led this unsettled and irritating life for twenty-five million years, half the time afloat, half the time aground, and always wondering what it was all for, they never suspecting, of course, that it was a preparation for man and had to be done just so or it wouldn't be any proper and harmonious place for him when he arrived.

And at last came the monkey, and anybody could see that man wasn't far off, now. And in truth that was so. The monkey went on developing for close upon five million years, and then turned into a man—to all appearances.

Such is the history of it. Man has been here 32,000 years. That it took a hundred million years to prepare the world for him is proof that that is what it was done for. I suppose it is. I dunno. If the Eiffel Tower were now representing the world's age, the skin of paint on the pinnacle-knob at its summit would represent man's share of that age; and anybody would perceive that that skin was what the tower was built for. I reckon they would, I dunno.[44]

As Twain's sarcasm suggests, *Homo sapiens* does not represent the pinnacle of progress. Those who act as spokesmen for the special interests of human beings fail to see how interdependent life on Earth really is. You cannot see evolutionary history in a balanced manner if you look at it only as a preparation for humans. Eighty percent of life's history was microbial. We are recombinations of the metabolic processes of oxygen-using and other forms of bacteria that appeared during the accumulation of atmospheric oxygen some 2,000 million years ago. Remnants of bacteria, with DNA identified by molecular biologists as extremely similar to free-living bacterial DNA, divide as mitochondria within the nucleated cells of people as these lines are read.

Human beings are not particularly special, apart, or alone. A biological extension of the Copernican view that we are not at the center of the universe deprives us also of our place as the dominant form of life on the planet. It may be a blow to our collective ego, but we are not masters of life perched on the final rung of an evolutionary ladder. Ours is a permutation of the wisdom of the biosphere. We did not invent genetic engineering, we insinuated ourselves into the life cycles of bacteria, which have been directly trading and copying genes on their own for some time now. We did not "invent" agriculture or locomotion on horseback, we became involved in the life cycles of plants and animals, whose numbers increased in tandem with ours.

In the same way, the much vaunted accomplishments of technology, from writing in southwestern Asia over 10,000 years ago to the modern microchip, are not *our* property. They came from the biosphere—from the interconnected environment of *all* life—and eventually, even if they should have to evolve again, they belong not to us but to it. In accord with our theory of the spirochetal nature of intellect, high technology is not really ours but planetary in nature. We have been separating ourselves from the rest of life, incubating forms of organization that are ultimately bigger and richer than we. We have done well separating ourselves from and exploiting other organisms, but it seems unlikely such a situation can last. The reality and recurrence of symbiosis in evolution suggests that we are still in an invasive, "parasitic" stage and that we must slow down, share, and reunite ourselves with other beings if we are to achieve evolutionary longevity.

The story of the rise and rapid spread of human beings over the face of the planet is hardly a story of conquest. Like wealthy and spoiled heirs, we have directly inherited genetic riches from animals which survived the earth's most massive extinctions, including the famous Cretaceous events of 66 million years ago, which destroyed not only the dinosaurs but many families of mammals and marine plankton; and the even more devastating Permotriassic extinctions of 245 million years ago, which killed off fully 52 percent of all families of life on Earth at the time (as compared to 11 percent for the Cretaceous). Various theories have been offered to explain these extinctions. One is the breakup of the single world continent Pangaea into two continents at the time of the Permotriassic extinctions. Another is the landing of a massive meteorite at the time of the Cretaceous catastrophe.

During the last 500 million years, there have been four mass extinctions even more devastating than the one that killed off the dinosaurs. Statistically, mass extinctions have been claimed to appear every 26 million years or so. This has suggested to some, especially to astronomer Richard Muller of the University of California at Berkeley, the existence of "Nemesis," a hypothetical sister star to the sun. Nemesis, it is suggested, cyclically pulls comets out of orbit in the Oort cloud and sends them whirling in toward the sun, some of them impacting the Earth and destroying vulnerable parts of the biosphere. (The Oort cloud, named after Dutch astronomer Jan Oort, is a belt of comets and cosmic debris beyond Pluto. It may represent material left from the origin of the solar system.) Others have suggested that regular vertical movement of the solar system through the galactic plane of our spiral galaxy accounts for the extinctions. The sun takes about 250 million years to circle the center of the Milky Way, but on the way it bobs up and down. Going through dense regions of the sky would periodically loose barrages of comets from the same Oort cloud. Whatever the reasons, the progeny of the survivors of such periodic devastations inherited the earth. Like the phoenix rising from its own ashes, DNA regenerated into new long-lasting forms. We can trace our history, but we should remember that it is no more necessary or noble than that of any other species alive today.

The first vertebrate animals, relatively large creatures with calcium phosphate backbones and a brain case shielding an elaborate nervous system, may have evolved about 510 million years ago from the tadpolelike larvae of invertebrate animals. Judging by the similarity in foetal development among so many animals, and from bone fragments in the late Cambrian record, one theory is that the precocious young of some invertebrates became capable of reproduction without first undergoing metamorphosis into adult form. These ancestral animals, called chordates, began with spinal columns but no bones. Like the Cambrian shelly forms, experiments with stockpiling calcium led to divergent forms, such as the first fishy animals, with external armor but lacking jaws. These early, jawless fish, most about the size of a ruler, fed by sucking in water and whatever it contained.

The periodic loss of coastline due to geological forces literally left many fish out of water and panting on tropical shores. As water in tide pools, lakes, and freshwater ponds depleted, nearly all the primeval aquatic animals died out. One exception were creatures like Australian and African lungfish; they are thought to have survived. Lungfish today breathe both highly oxygenated water and the air itself. Lungs are thought to have evolved from buoyant gas bags, called swim bladders, formed by balloons of gut. Used in many modern fishes today to move up and down in the water, the modified gut pocketings of some ancestral fish lent themselves to oxygen use when early fish were stranded in muds or dry land. When

TABLE 2
HUMAN CLASSIFICATION BOX

The classification of people traced from first to evolve (most inclusive taxon) to most recently evolved (least inclusive taxon)

Taxon (grouping)	Comment on Taxon	When?*
Kingdom Animalia	develop from a blastula	750
Phylum Chordata	hollow nervous system in back, spinal cord and brain with fluid, gill slits	450
Class Mammalia	hair and mammary glands modified from sweat glands, give milk to infants	200
Order Primates†	anatomically unspecialized mammals including monkeys, apes, and bushbabies	60
Family Hominidae‡	apemen and manapes	4
Genus *Homo*	all extinct but us; *Homo erectus* probably became us	0.5
Species *sapiens*	artists, poets, food gatherers, big game hunters	0.01

* in millions of years ago

† pronounced "pry-mah-tees"

‡ These are the manlike apes, including australopithecines like Lucy (p. 216), that have been around for some fifteen million years. Most biologists feel living apes and people are so similar that we all should be placed in the same family: Pongidae/Hominidae

lakes and rivers seasonally dry up in Australia and Africa a substance is produced in the brain of the lungfish that lowers its activity as if it were hibernating. While other fish die in the muddy river beds, lungfish of the genus *Dipnoi* routinely survive to live another season. But *Dipnoi* were inconsequential in vertebrate evolution—they are confined to drying rivers in isolated parts of the globe. They were not direct antecedents to our fishy ancestors that started to hobble onto land.

Another lineage of fish (which also had lungs but are not called lungfish), the fleshy-finned *Crossopterygii,* originated some 400 million years ago in the Devonian period. They probably evolved to be the first amphibians. *Eusthenopteron* is a fossil genus of this lineage. With fishlike spines and stubby fins, *Eusthenopteron* was near the main line of animals that crawled along the shores. With jaws—a unique fish contribution to the fantastic later development of all land vertebrates—and a froglike head, *Eusthenopteron* looks like a cross between amphibians and fish.

Forms similar to *Eusthenopteron* are found in the fossil record until Triassic times. The ancestors of amphibians, reptiles, birds, and mammals all have lungs, which are thought to be similar to modified gills that could withstand waterlessness for short stretches. By the time of the Permotriassic horrors, pools with fish in them had already dried up many times. While some believe the ancestors to frogs, salamanders, and newts had already come onto land, others suggest that modern frogs and salamanders are sufficiently different from the early amphibians to indicate a separate origin from fishes, perhaps some 240 million years ago.

The early amphibians spanned the late Devonian to the end of Triassic times, 345–195 million years ago, peaking 310 million years ago as lords of the coal swamps. *Ichthyostega,* recreated from fossils found in Greenland, was one of the first true amphibians. With fleshy feet, a remnant tail of fish scales, and an amphibian head, *Ichthyostega* has a skeletal structure related to that of air-breathing fishes. In the late Paleozoic (400–245 million years ago), in swampy coal forests, among club moss trees and giant seed ferns, amphibians radiated exotic and numerous species, of which little really familiar remains besides frogs.

By Triassic times, however, the evolution of reptiles was underway. By 245 million years ago, reptiles had displaced amphibians from land, and by 195 million years ago, at the end of the Triassic, from the water as well. The reptiles had powerful jaws, desiccation-resistant skin, and, most importantly, a new sort of egg. They enclosed and encapsulated the aquatic environment of their amphibian ancestors. Instead of the many, small, jelly-wet eggs of amphibians, reptile females laid fewer, larger, and tougher ones. Inside these larger eggs reptilian forms developed fully prepared for life on land. This amniote egg with its large allotment of yolk prepared reptiles for land living. A recreation of the ancient ocean environment, the reptilian egg (and later the mammalian womb) recreated or took with it the ancient time of water life. The shelly exterior of reptilian eggs retained water yet still permitted air to exchange. Today we still have many types of lizards, the main lineage of which evolved 240 million years ago. Among these are the legless lizards or snakes, the chief family of which evolved only 30 million years ago.

Amphibians never freed themselves from water. Even today they fertilize their eggs and develop through tadpole stages in lakes, streams, and puddles. By contrast, the early phases of embryo development in reptiles all took place within the watery environment of the fertilized egg. This encapsulation was a brilliant evolutionary innovation, comparable to that of the seed of Paleozoic seed ferns. In reptiles, the typical fins of the fish ancestors modified. Today the comparative anatomy of lizard feet, horse hooves, and human hands shows that such limbs are modified fins, preserving an internal bony structure beneath common to all four-limbed beings with backbones.

Another major change accompanying the transition from living in a liquid to living in a gas was the development of keratin. Keratin, the characteristic protein of reptilian skin and mammalian hair, made it possible for hatched reptiles to withstand desiccation. Keratin may be one of the few metabolic innovations that did not come directly from bacteria and their consortia.

The first amphibian-derived reptiles, called stem reptiles, can be typified by *Seymouria*. *Seymouria* is a fossil genus midway between amphibians and reptiles. Recreated from fossils found in west Texas, *Seymouria* is an animal that may be directly ancestral to humans.

From such stem reptiles—the first vertebrates adapted to land—came fantastic evolutionary expansions called "adaptive radiations." The Mesozoic (245–66 million years ago) reptiles mated and laid eggs all over the green, swampy land. Stem reptiles leading to mammals had already appeared by Permian times (290–245 million years ago). An evolutionary question mark with no ensurance of any future success, the first mammallike reptiles were as inconspicuous a life form in their time as hedgehogs are in ours.

About 216 million years ago another line of reptiles evolved into some of those "terrible lizards," the dinosaurs. Like the fish before them and the birds after them the dinosaurs had backbones and laid eggs; they presumably had the usual whip-tailed 9 + 2 sperm, and 9 + 2 rod cells in the light-sensitive layers of the retina of their eyes. They too came from the microcosm. In some species corrugated back sails may have acted as passive solar collectors to regulate body temperature. Some species flapped wide skins and flew. Some swam. Some female dinosaurs hatched their eggs inside their nourishing bodies. Through the cloacal opening these mothers bore live young. The part of our brain anatomy that appears not only in apes but in snakes and crocodiles—the medulla oblongata or brain stem—may to this day be part of our large reptilian heritage.

About two hundred million years after the Triassic period's adaptive radiation, which established the major reptile species—sea serpents and small running dinosaurs as well as more familiar sea turtles, snakes, and lizards—the dinosaurs all died. The Cretaceous extinction of the dinosaurs has always been cause for speculation. Some assumed they were too big and stupid, brutes whose time had come. Recent evidence suggests otherwise. There is good evidence that the sudden extinction of dinosaurs and many other forms of life 66 million years ago was extraterrestrially induced, caused by an incoming planetoid or planetoids from outer space.

Iridium, rare on Earth, is often found in meteorites. It is strangely abundant in sediments of the late Cretaceous period, especially those at the Cretaceous-Tertiary boundary that marks the death of so many species. The presence of a worldwide iridium layer in rocks of this age has been interpreted by the scientific father-son team of Luis and Walter Alvarez and their colleagues in California to mean that a giant meteorite about seven miles across hit our planet. They suggest that meteoritic dust caused a prolonged, worldwide blackout. With meteoritic dust reflecting heat and light into outer space, the earth would have cooled and photosynthesis would have dramatically decreased. The dead bodies of bacteria, protists, and plants accumulated. Nonphotosynthetic bacteria, protists, fungi, and animals starved. Production stood still. Extinction followed extinction. The huge dinosaurs succumbed.

This apocalyptic scenario may be rather melodramatic. Paleobotanist Leo Hickey, director of Yale University's Peabody Museum, points out that many species of land plants seem unaffected at the Cretaceous-Tertiary boundary. If large numbers of forest trees, shrubs, and herbs grew happily for millions of years, leaving fossils both before and after the boundary, how could there have been a persistent, worldwide duststorm and blackout? Yet foreign objects may have crash-landed on Earth, causing drastic alterations in weather patterns. Such impacts would have led to sudden alterations in temperature, light intensity, sea level, and so forth, with the consequent destruction of many, but not all, communities of organisms. If the meteor or meteors were large enough, elements in them poisonous to some forms of life may have directly entered the atmosphere. Plants tolerating dramatic seasonal changes may have been survivors.

Beginning about 210 million years ago, the mammals came into their own. Scurrying and prowling under the light of the moon, the oldest true mammals are thought to have been small creatures awake and active during the night. Cool air inhibits reptiles, unable to regulate their body temperatures very well. Mammals, however, do not become sluggish but more active in chilly surroundings. Muscle movement generates heat, which in mammals is controlled and used. And so from their inception, evolving from fishy looking creatures like *Seymouria* through mammal-reptiles like the "dog-toothed" *Cynognathus*, mammals were more fit to deal with the cold than the stem reptiles from which they presumably sprang. Although there were probably temperature-regulating reptiles, continuous activity in the cold has always been a mammalian trait. Up and alert through the night with sharp dilated eyes, mammals strayed from a daytime dependence on the heat of the sun and wandered northward and southward. As they crept away from both the warmth and the reptiles of the equatorial tropics, mammals evolved better means of insulation to maintain their internal temperatures. Perhaps some grew feathery; our ancestors certainly secreted the proteinaceous strands of skin cells we call hair— another use of the protein keratin, this time as a means of protection against cold air. This tough animal protein that evolved in reptiles is still used in making organic tools and weaponry, such as bird claws and rhinoceros horns.

As the mammals evolved from reptiles, females no longer put their fertilized eggs in holes in the ground or out in the open air. Instead they nourished their young inside their own bodies, in the heat of the maternal womb. After birth, their hungry young licked secretions from sweat glands on their mothers' bellies. These sweat glands were mammaries, and the nutritious sweat was the calcium-rich liquid, milk.

From their beginnings approximately 200 million years ago, both the mammals and birds were distinguished from their reptilian ancestors by special attention to and care for their young. Most reptiles have many offspring which they abandon after hatching. But modern-style birds (since 133 million years ago) and mammals feed and teach fewer, more vulnerable offspring. Unprecedented activity and parental commitment distinguished birds and mammals from the start.

The early primates at the start of our own Cenozoic Era (since 66 million years ago) were small, wily animals, quick to jump and evade predators. Some of these insect-eating mammals even clung to branches with the modified claws that in humans were to become fingernails and toenails. Hopping from branch to branch in the forest at night placed a premium on keen vision. These first primates were animals much like tree shrews and lemurs and are called prosimians, from *pro,* meaning prior, and *simian,* meaning monkeys. The prosimians are thus primates which predate the first monkeys. Presently they are primarily restricted to Madagascar and southeast Asia, to places not overrun with monkeys. Fossil prosimians are known from Asia and Africa.

All primates today except ourselves are vegetarians or insectivores. They eat nuts, berries, fruit, grasses, and insects. Some have specialized diets, like the lemurs—one kind shares its nest with bees while feeding on nectar, while another has a long, pointed, second fingernail used for digging out meals of termites and ants from logs. We humans are the only meat eaters among the primates. Carnivorousness is a habit that, judging by fossil teeth, we acquired late in our evolutionary history.

Taking refuge in trees where they could escape the notice of less agile animals, tree shrewlike primates not unlike some of the small placental mammals in Madagascar evolved what is called orbital convergence—a gradual centering of the eyes from the sides to the front of the head. This optical centering was crucial to three-dimensional vision, which in turn was indispensable to the judgment of distances in trees. Primates unable to assess distance with consistency tended to plummet from high branches to the forest floor below. Our ability to judge heights, our propensity to construct and tinker, perhaps even our propensity to live in multilevel apartment buildings, owes a debt to orbital convergence, the movement of the eyes from the sides to the front of the head over many generations.

Another mark of the arboreality or treegoing expertise of some early members of the Order Primates was the development of clinging hands and feet with flattened nails and opposable thumbs. Prehensile or grasping tails, found only in the monkeys of the Americas, is a later primate adaptation to living in trees.

Today in human foetal development there appears a vestigial tail remarkably unsuited to any human need. Occasionally a child is even born with a little wispy tail. Although the appearance of newborns with tails was often attributed to the Devil, even normal infants show traits that hark back to our arboreal beginnings. Such useless appendages were one of the things that impressed Charles Darwin with the method in nature's madness. Rudimentary organs, as they are called, are parts which serve no function in present survival. They make little sense except as holdovers from an earlier evolutionary time when they were needed in survival. (Even biparental sex, as we have seen, makes no sense except as an odd inheritance from the life cycles of our ancestral protists.) But evolutionary rudiments need not be structures like the appendix or tailbone; they may also be *processes*. Soon after birth all infants naturally make tiny fists around proffered fingers. This universal grasping response must have saved ancestral primates from falling by allowing them to hold mother's fur and the nearest branches. Similarly, the feeling of falling that occasionally jars the would-be sleeper to sudden wakefulness may be a psychological response held over from living in trees. Sleeping in trees is hazardous. In both the instinctive activities of infants and the psychological propensities of adults there lie rudiments and reminders of our prosimian ancestors. No wonder children are scared of monsters when left alone in the dark, despite the obvious safety of their rooms. Once upon a time, such deep emotions had an unquestionable survival value: helpless children were especially delicious to woodland predators.

The first primates then were mostly tree-climbing cowards. They ran and hid and clung to branches in the starry night. The running and hiding, the clinging to branches and active night life was supplemented by another element of primate cowardice which was ultimately of immense importance to primate evolution: the incipient tendency toward cooperative social behavior. Making frequent loud noises is a way to stave off enemies; as a primate survival tactic noise-making preceded true vocal communication. Today, like wild dogs, chattering baboons will team up on single predators in the open savannah of Africa. Like cells and social insects before them, our ancestors united in crowds. En masse they dared do what no individual would.

At various times and places when nuts and fruit were in short supply our ancestors descended to the ground. Keeping guard over the tall grass required upright posture and fast glances in all directions. Baboons do this today, after which they quickly return to a crouched posture. Such reconnaissance was strongly rewarded: those animals with their heads up ultimately freed their hands to dig for roots, to throw rocks and wield sticks as weapons, to build and explore. Feet became flatter as simian fingers manipulated. Today many monkeys and apes use their fingers to tease young grass from the ground and eat its tender tips or poke around for insects with sticks. One caged orangutan in the Central Park Zoo reportedly throws feces at onlookers who toss peanuts or popcorn beyond his reach. We clap and shake hands to show appreciation. The potentialities of hands would never have occurred were they not freed from the redundant tasks of locomotion.

Just as a natural excess in the production of calcium and fast-moving symbionts gave natural selection the raw materials with which to mold skeletons and carry on mitosis, so upright primates found themselves with a redundancy of clasping extremities. The provision of extra reproducible elements that are freed to perform new functions can be classified as *redundant innovation*. Redundant innovation occurs from the molecular to the social level. Molecular redundant innovation occurred when plants cast away waste alkaloids that took on new active roles as feeding deterrents, and when animals evolved hormone glands from preexistent biochemicals used casually in cell communication by the genetic systems of their ancestors. Social redundant innovation occurred when societies of termites and bees diverged into workers, soldiers, and queens. Society undergoes specialization as have the cells of the clones of the members which make it up; but this specialization which confers survival value could not have occurred without the prior superabundance or redundancy of reproducing elements. The use of hands for tasks other than locomotion, which at first was occasional and dispensable, became absolutely necessary as apemen emerged.

One widely accepted account of human evolution suggests that people are developmentally retarded apes. Technically known as *neoteny*, from etymological roots meaning "keeping the new," the retention of youthful traits into adulthood may have been of crucial importance in human origins. The length of time needed for human beings to pass from infancy into childhood, and to reach puberty and adulthood, is longer than maturation in any apes. The times of dental eruption from the gums and closure of bone sutures in the cranium after birth—calcification events which have Paleozoic (580–245 million years ago) precursors—are retarded in humans relative to chimpanzees, orangutans, or gorillas. The human baby is thus a sort of extra-uterine ape fetus, while human adults are similar to apes that have never grown up. But the humbling fact that we are retarded apes has its compensations. It is a sort of double negative that flips into its opposite. To be stupid at being stupid, or slow at being slow, can be put to advantage. Delayed development gives us time to learn to be smart.

From the ape's point of view we are born too soon, before

we are ready. After birth, the weight of the average human brain nearly trebles in the first two years, from 350 to 1,000 grams. While chimpanzees and orangutans are wallowing in intra-uterine bliss, human beings can get a literal "head start." We are exposed and molded by the influences of the outside world while we are still malleable and vulnerable. Unlike antelopes born on the run or turtle babies who never know their parents—unlike any animals whose young develop fully formed and prepared for the outside world—our infants are half-formed and utterly helpless. Our completely dependent infants require a long period of maturation in association with adults. The great and often repeated step of protective growth in the parental body has been turned round. Why did evolution move in the opposite direction with humans?

From the fossil record we now know that our bipedal posture, our walking on two legs, preceded the development of our swelled heads. But apes with small heads and upright gaits were still not humans. Some apes may have given birth to infants prematurely. The births were easier, since the heads of the infants were smaller. Premature ape infants could have been imprinted with experiences from the harsh outside world at earlier ages, giving them more time to learn to modify their behavior. Experienced ape children can be presumed to have learned better and at a younger age adult survival tricks. Since premature birth is in part due to a genetic predisposition, ape couples who themselves retained youthful traits into adulthood can be assumed to have given birth to more prematurely-born and neotenous children than their peers.

From this genetic change in timing came crucial selection pressure for more premature birth, more mutant lines of premature babies. The ape babies, born relatively hairless and underdeveloped, either died or were protected and thereby educated in the outside world. Some grew into relatively hairless adolescents who retained their childlike features well into adulthood. They gave birth to babies with bigger brains, retarded apes more intelligent than any ape. This same scenario is further enhanced by adding the element of sexual selection: when males chose spouses reminiscent of the prehistoric "fertility-goddess" figurines found in Europe—women with wider hips, larger buttocks, and correspondingly larger birth canals—they could, hypothetically, have sired still larger-brained infants.

The advantage of the early birth of highly impressionable young infants had its disadvantages as well. Such minds inhabited pathetically defenseless and tiny bodies. Yet out of infant helplessness—and this is strange—came family, came civilization. Females were forced to alter their strategies of sexual selection. They desired a new kind of apeman. Young mothers attached themselves to males who would care for them while they cared for their fragile, unformed children. Women chose large-brained men and large-brained men chose wide-hipped women; feedback loops started. Intelligence began to beget intelligence. More specifically, it is thought that by throwing rocks, and stunning or killing small prey, early humans were catapulted into a new evolutionary niche. The psychomotor skills necessary to plot the trajectories of projectiles, to kill at a distance, were dependent on an increase in the size of the left hemisphere of the brain. Language abilities (which have been associated with the left side of the brain, which controls the right hand) may have fortuitously accompanied such an increase in brain size. Indeed, William H. Calvin, a neurobiologist at the University of Washington in Seattle, even suggests that our predominant right-handedness comes from a time when mothers, hunting alone, and clutching their infants to their left breast so as to quiet them with the comforting sound of the human heart (which they sorely missed due to their early ejection from the womb), threw rocks at small animals with their free right hand. So, just as computers were first developed to calculate the trajectories of enemy airplanes, so both the tendency (unknown in apes) of humans to be right-handed and our speech and writing may be due to the increase in parts of our brain used for sequencing functions originally reinforced by our maternal ancestors' success in throwing stones at small prey.

Finally, in the post-ape world of retarded development, females lost estrus; they no longer went into "heat" at specific times of the year. Such women were considered more attractive. Today, all monkey and ape females enter estrus. The genitalia of some monkeys, baboons, and other primates turn colors in response to hormonal changes designed to attract mates. But human females are different: they may feel sexy at any time and do not go into heat per se at all. Males who devoted time to their families may have relinquished random sex during a change of social organization. Since females were increasingly sexually receptive the whole year round, there would have been less need for males to roam about in search of sexual partners. An apeman providing for an apewoman and their needy infants was kept busy with only a female or two from whom he could demand sexual receptivity at virtually any time. The family that laid together stayed together.

Nomadic bands and villages may have arisen as males, females, and children bonded into communities which could serve as an overarching protector of the new family unit. Perhaps the "Mother Earth" idea of early—and indeed modern—cultures comes from the exposure of infants to the outside world at a time when they are more prepared to be in the womb. Psychologically, land replaces mother, and the world becomes a sort of second womb.

Another example of the importance of neoteny in human

origins has to do with the fact that almost all mammals lose the ability to digest milk sugar after infancy. Milk sugar or lactose is still harmful to most adult humans. But large groups of adult human milk-drinkers, including the Semitic, Uralic, and Indo-European peoples, exist. They exist through a mutation which spread to populations in northern Mesopotamia some 8,000 years ago. The rise of mutant adults tolerant of milk paved the way for a livestock revolution more profound than the overthrow of any despot: the milking of cows, goats, and sheep, the use of draft oxen and plows, and, later, the taming of horses. Tolerating milk gave adults a source of animal protein that did not require killing prey, and led to nutritious milk products such as cheese, yogurt, whey, and butter. Another bonus was that since milk contains vitamin D, which transports calcium, it ameliorates vitamin D deficiency, which causes rickets, a disease of the bones. Melanin, the dark skin pigment, also is helpful in preventing rickets. Indeed, the adult milk drinkers by and large had fair skins. As Nigel Calder writes: "With the revolution of cow and plow came the social divisions that characterize Eurasian cultures, as between rich and poor, lords and serfs, and men and women. War became commonplace, and warriors set themselves up as kings."[45] Although human beings had already evolved some four million years prior to the spread of these first cowboys, we mention it because the mutation that allowed some people to digest lactose after infancy was also a form of neoteny. Delays in genetic timing that slowed aspects of primate developmental processes may have brought human beings to the very dawn of traditional history. Elaborate clay tablets, adorned with many symbolic drawings and signs, appeared in Mesopotamian cities some 5,500 years ago.

But we must be on guard when we ascend from the microbial world to examine our own history. Any summary statement will be a generalization, and generalizations of generalizations usually become better stories in direct proportion to becoming worse histories. While we have stressed the "retarded ape" or neotenous idea of human development, by itself this presents an oversimplified picture; it should not be mistaken for more than one aspect of a multidimensional story. It is common to picture the evolution of humanity as if it were the epic transformation of a single evolutionary hero. A lungfish thrashing about in the water, developing amphibious limbs to crawl upon land, and eventually standing up in triumphant humanity is suspiciously close to the real-life experience not of evolving populations but of a single human being. It is *we* who emerge from the birth canal coughing for air and later learn to crawl and walk. So, too, early man, depicted in our evolutionary hero-worship, is a creature who stumbles out of his primeval cave with club in hand. He stands upright and stretches before the full light of modernity. But the hairy "caveman" coming out of the great dark cave of prehistory is amazingly simplistic. A cultural myth reinforced by cartoons and comic books, it is more a queer amalgam of dim childhood memories and archeological storytelling than a true account of the various and sundry real-life appearances and extinctions of ancestral apes through time.

The true complexity of our ancestry is impossible to imagine. Of course, most of it was not only pre-ape, it was microbial. But thirty million generations of highly varying mammalian populations cannot be abstracted into a concise pictorial or literary image without severe distortion. Still, the tendency is to connect the dots, to make single, familiar, anthropomorphic images. Make-believe evolutionary heroes relieve us from the strain of visualizing the true multiplicity of populations, of beings nonhuman and strange.

We are primates and primates are tropical animals: we hate and fear the cold. Since primates are most diverse and abundant in the Old-World tropics, and since all fossil apemen have been found in Africa, Europe, and Asia, everyone agrees that the tropics of the Old World was the original home of our primate ancestors. The only apes in Antarctica, Australia, North or South America are humans or their captives in zoos.

Objective scholars, if they were whales or dolphins, would place humans, chimpanzees, and orangutans in the same taxonomic group. There is no physiological basis for the classification of human beings into their own family (Hominidae—the manapes and apemen), apart from that of the great apes (Pongidae—the gibbons, siamangs, gorillas, chimps, and orangutans). Indeed, an extraterrestrial anatomist would not hesitate to put us together with the apes in the same subfamily or even genus. Human beings and chimps are far more alike than any two arbitrarily chosen genera of beetles. Nonetheless, animals that walk upright with their hands dangling free are aggrandizingly defined as hominids (family Hominidae), not apes (family Pongidae). The only hominid left alive today is us.

Yet many fragments of distinctly different hominids have been found in the fossil record. Two distinct kinds of hominids have been identified which spark the anthropologist's imagination. The two sorts of animal in that gray area straddling man and ape are australopithecines (apemen) and homos (manapes). Several species of the genus *Australopithecus* endured until about a half a million years ago, when they became extinct. Some australopithecines may have been our ancestor; others probably were not.

In the absence of any evidence to explain the extinction of the australopithecines, some anthropologists have speculated that *Homo* may have caused his dumber cousins to become the first members of humanity's original, unwritten list of endangered species. The end of the larger, vegetarian hominids may have been hastened by massacres or, in the case of the apeman *Australopithecus robustus,* by an obsequious overdependence on the earliest true humans. Members of this species may even have been enslaved by members of our genus, *Homo.* The lumbering *Australopithecus robustus* may have followed our ancestors about, feeding on their offal. It is even possible, according to the scenario developed by Benjamin Blumenburg of Leslie College and Neil Todd of Boston University, that the big, dim-witted vegetarians became extinct when our ancestors replaced them with early dogs. Evolved from wolves, the dogs would have hunted better, guarded better, sensed odors far better, and, being smaller, been far easier and less expensive to feed than any camp-following *A. robustus* apes.

Three prehistoric species are identified as members of our genus, *Homo.* In order of evolutionary appearance they are *H. erectus, H. habilis,* and the older subspecies of *H. sapiens. Homo habilis* ("handy man") was an ancestral species of human who apparently began using neatly chipped stone tools. *Homo erectus* ("upright man") may have been the first to use fire. *Homo sapiens* ("wise man") is our species. *Homo sapiens* is divided into at least two subspecies, *Homo sapiens neanderthalensis* (Neanderthalers) and *Homo sapiens sapiens* (us). Most anthropologists agree that all but *Homo sapiens sapiens* are extinct.

Hominids appeared on the scene perhaps two million years ago. Reconstructed from fewer than a thousand well-identified fossil remains, these extinct hominids constituted the primate line leading to us but not to the gibbons, gorillas, chimps, and orangutans. The hominids, both the homos that became us and the australopithecines that either became homos or died out, were two things above all: tropical and African.

Not only fossil primates but fossil apemen and tools have been dug up almost exclusively from tropical and semitropical regions in the Old World. The earliest stones worked by human or semihuman hands have been found in Africa. Humans are more similar in their structure, protein chemistry, facial expressions, and social behavior to the tropical apes of Africa than to any other living species of animals. Furthermore, we retain adaptations to living in hot climates. We can dissipate lots of heat, 400 to 500 kilocalories per hour, mostly by the evaporation of sweat.

Unlike the African apes, however, we sweat all over our bodies—some one to two quarts an hour can be evaporated by about two million little sweat glands. Our blood also functions as a coolant. A large proportion travels to below the surface of the skin. As we work in hot climates the volume of blood increases by retention of water and salt. Tiny blood vessels carry the blood to the undersurface of the skin and thus serve as conduits of heat to cool us. Since these adaptations to working in the heat are shared by all humans it is logical to argue that they were shared by our immediate ancestors after they diverged from some common line to us and other living apes.

While the gatherer-hunter story-teller appeared in the evolutionary drama relatively recently—fewer than a hundred thousand years ago—our species in transition shows continuity with African australopithecines. We are probably descended from mothers such as "Lucy." Named by her finders after the Beatles' song "Lucy in the Sky with Diamonds," which was playing at the archeological dig near the time of her discovery, Lucy was a north African apewoman whose fossil remains date back over three million years. Her amazingly complete fossil skeleton was found in the Afar triangle near Hadar, Ethiopia. She was an upright walking and running animal about three-and-a-half feet tall. Although she had the pelvis of a petite woman, Lucy's face was that of a chimp.

Partial skeletons and teeth of several dozen individuals from the same area have been found. Lucy, according to one of her discoverers, Donald Johanson, is the prototype of the ancestral hominid, *Australopithecus afarensis*. Indeed ape-people like Lucy and her relatives, namely us, may have had common chimplike ancestors only four or so million years ago.

The new science of molecular evolution helps us estimate the timing of evolution. The data from large molecules allows us to count the changes involved in the divergence of two lineages. For example, amino acid analyses of common proteins such as chimp hemoglobin and human hemoglobin show few molecular changes. From the number of differences the time of divergence of ancestral forms can be estimated. Human hemoglobin and ape hemoglobin are remarkably similar. Using the macromolecular scale developed for other animals, some biologists, such as Alan Wilson at the University of California, Berkeley, have concluded that the African chimp lineage diverged from ours only about 4 million years ago, not the 15–20 million years that was previously thought.

Early humans, whose ancestors lived in the Pliocene Epoch from 7–2 million years ago, came into their own in the succeeding Pleistocene Epoch. It was during the Pleistocene that the famous ice ages occurred. At the time of their greatest extension, sheets of ice covered central Europe, Asia, and Siberia. In England ice blocks spread past the river Thames, freezing up water from the ocean. In North America glaciers covered New England and reached what is today central Indiana, Illinois, and southern Ohio. Early man had spread like an epidemic from his tropical homeland in times of relative warmth. But, far from home, he was caught off guard and tested by bitter cold and snows to which he was not originally accustomed. In interglacial periods warmth visited again; but melting ice caps flooded, robbing him of coastal lands. Giant Pleistocene animals such as hippopotamuses, mastodons, rhinoceroses, and saber-toothed tigers were warmth-loving mammals that, unlike man, did not pass the last Pleistocene test of frost, ice, and freezing snows. These animals, stalked as prey by hungry, running people, were replaced by new fauna. Musk-oxen, wooly mammoths, wooly rhinoceroses, lemmings, bison, auroch, caribou (reindeer), and other cold-weather beasts were forced into existence by the persistent ice. Our heritage is tropical, but we are children of the ice ages.

The dates of the ice ages are not precisely agreed upon. From the earliest to most recent the ice ages are named the Gunz (which started about 700,000 years ago), the Mindel, the Riss, and the Wurm. These European time periods of maximum ice extension roughly correspond to the Nebraskan, Illinoisian, Kansan, and Wisconsan ice ages in North America. The interglacial or warm periods between the ice ages are known as the Gunz/Mindel, the Mindel/Riss, and the Riss/Wurm. The last ice age was probably at its maximum a mere 18,000 years ago. We are still recovering from this worldwide winter.

The most impressive record of continuous habitation by humans is to be found today in the dry sunlight of East Africa at the Olduvai gorge. Since about two million years ago (at the beginning of the Pleistocene and before the more recent ice ages), this equatorial place has been choice real estate.

In what is today Tanzania, the great gorge at Olduvai has been continuously revealing fossil remains. We can picture the venerated Olduvai school system with its courses in elementary, intermediate, and advanced toolmaking. The fashioning of stone tools is the oldest human industry, the seed of technology. Hammerstones (adzes thought to be aids in making other tools) and anvils (rocks used to hold other worked stones while they were being flaked into form) have been dug up. Unmodified flakes of rock and other fragments, fancifully called "debitage," have also been found in Olduvai. They were probably used to cut meat or wood. "Manuports," hand-carried volcanic rocks, although unmodified, are considered artefacts, too, because they are found removed, sometimes at great distances, from their sites of origin.

In some lucky places in the gorge entire scenes can be reconstructed. In the lower part of "Bed II" a skeleton of *Deinotherium,* an extinct elephant, was discovered, caught in a bog. There is evidence that the mammoth got one of its great feet stuck; the hominids severed the creature's head, crushed its skull, and left one of their choppers sticking into the bones of the pelvis. No doubt they feasted on the remains of the poor trapped beast. There are, however, no human skeletal remains; the hunters apparently avoided sinking into the bog themselves.

Whether these hominids were apemen of the *Australopithecus* types or manapes of the *Homo* sort is not known. But they must have been social animals. Clearly they were brave and tough. They were far more versatile in employment than today's bureaucrat or plumber. But though less specialized, these Bed II hunters were like the worker of today in that they were dependent on their fellows for food. They did not feast alone. No single man, however heroic, took on that mammoth himself. Since they hunted together, they must have habitually shared their food. Food sharing is considered to have been catalytic in the development of human culture and civilization. In anthropological story-telling, the sharing of food has been contrasted with the development of stone axes and spearheads. Cultural critic and historian William

Irwin Thompson has summarized the differences between these two perspectives:

> One group of anthropologists say that it was the weapon that made us human. As we took stone tools in hand, we fell from animal innocence and turned to confront nature armed with the first technology. We hurled ourselves up into a new technological level, and everything that was left behind was the primitive. And so, in this view, human culture is created and determined by technology, and the most basic technology of all is the weapon. . . . Fortunately, there is another scientific vision of human origins, one associated with the work of the anthropologist Glynn Isaacs. Here the theory has it that the protohominids carried their food to a safe shelter, and there in an act of community definition, they shared their food. From this perspective the primary act of human culture is food-sharing. And here we can see that in the sharing of food together we most truly perform our basic humanity.[46]

The Olduvai hominids ate well and rolled around with their women at night. Certainly these women bore healthy children. At least some of these children learned from their parents to make and use the tools. And these hominids must have shared not only their food but their hunting plans. Those who remembered the plans and performed their jobs as planned probably continued to eat well. Those who planned and shared had children who ate well. The Olduvai probably passed their evolutionary examinations, leaving more children than other, less socially coordinated bands of manapes.

Modern civilization is an extension of dexterity and animal intelligence which developed in our ape ancestors. The ice-imposed socialization of early people was a harsh and unrelenting process. From crafting and sharing natural objects to hunting and donning the furs of cave bears and mountain cats, people have learned to outsmart large, threatening mammals. Much of the cohesiveness of the clan, the running chases of great elephantine beasts across the primeval plain have been preserved in modern cultures. Metamorphosing, these highly successful survival strategies have modern corollaries in team sports and war. In football the hunt seems reduced to the symbolic act of groups of men chasing an object made of animal hide. The ball is hurled through the air, a symbolic spear making its mark. So too, the tribal activity of war has not diminished but expanded. Our jabbering, gesturing ancestors hunted major species of large mammals to extinction. Today the momentum of big-game hunting has pushed our species to the brink of self-extinction. As Thompson writes "The technologist can turn upon traditional humanity to say: 'We are the highest, the most advanced; you are simply the sloughed-off remains of an old animal nature.' For these people the arms race is neither a necessary evil nor a peculiar pathology; it is the driving force of human evolution itself."[47] Actually, human evolution, like all evolution, had both aspects, sharing and slaughtering, competition and cooperation.

Language is a tool as fearsome as the sharpest point of flaked obsidian. We can envision the accidental and slowly reinforced development of language among early peoples. Sounds of vocal pleasure would be emitted by members of the hunt warily eating around a night fire. These sounds, repeated, became names representative of things. For instance, the word for "mammoth" might arise as one of the killers of the beast smacked his lips and grunted approval. The word would be recalled and associated with sights, sounds, and smells by those who participated in the feast, the hunt or, still better, both.

Out of this grew the development of a primordial vocabulary. Any vocabulary is better than none and has survival value. Those who shared the most words in common could communicate and organize the best in hunts—not only of furry quadrupedal animals but of each other. Why our ancestors developed a protolarynx, later to become the instrument of speech, is not known. But this was part of the physical baggage of the successful hominids, prerequisite to the deadly tool of verbal communication.

Scrawls were made upon the walls in animal blood. These primitive graffiti, a natural celebration of the kill, were at first physically associated with the kill; but finger painting eventually evolved to be representational. Sympathetic magic became symbolism. Juice of crushed red berries signified blood. Configurations of pigments became animals in the hunt. With the manipulation of symbols in various sequences people began to explore the possibilities of potential realities.

Over centuries the crude representations were clarified and streamlined. Some ritualistic marks became the hieroglyphs (picture symbols) and ideograms (picture symbols with sound correlates) of Egyptian and eastern civilizations. In rebuses the ordering of figures represents spoken sounds in addition to objects in nature. The more abstract symbolism of modern alphabetic languages replaced many rebus languages in time. Warning cries, communicative grunting, and meaningful scrawls enhanced the survival of the communicators. Common symbols bound together human communities around totem animals, religious icons, and other bits of abstracted nature. The Egyptians had a god called the Plan Maker, and it is reasonable to assume that hunter-gatherers were sketching maps and plotting the movement of the planets and stars as early as 40,000 years ago.

People had explored and settled the Old World by about 30,000 years ago. In Algeria, at a place called the "Afalou-Bou-Rhummel Rock Shelter 34," skeletons have been found near Mousterian tools, made by a Franco-German Neanderthal culture. The tools are made from pieces of obsidian, a natural black volcanic glass. Flaked into hand axes and tooth-edged spearheads, they are considered Paleolithic or old stone age. In Boskop, South Africa, pieces of a human skeleton, a jaw, and some parts of limb bones have been found, along with tools.

At Predmost in Moravia the direct remains of about thirty people have been unearthed. These burials were intentional; horn and bone implements were left to accompany the dead. In southern France and northern Spain, on the back walls of damp, inaccessible limestone caves are pictures of demigod wizards and halfhuman animals, testaments to cultures full of art and mystery.

Perhaps the most famous cave paintings are those of the Altamira caves in Spain and the Lascaux caves in France. Taken together the western European caves range in age from nearly 40,000 to 10,000 years old. The paintings vary from stick-figured hunters to elegantly portrayed extinct large beasts. Some show headless bears and bovine brutes. To arrive at such paintings the painters and spectators crept under low cave walls and twisted along cluttered paths, presumably carrying torches or some source of light. The three French brothers after whom their father named the Trois Frères cave had to spelunk along an underwater river before they made their enchanting discovery.

At the far walls' ends explorers find handprints of the ancient artists. On ceilings modern people gaze at eyes painted with zinc oxide pigments thousands of years ago. Such paintings alone clearly mark the presence of modern *Homo sapiens* on earth. Only people paint, only people plan expeditions to the rear ends of damp, dark caves in ceremony. Only people bury their dead with pomp. The search for the historical ancestor of man is the search for the story-teller and the artist.

The so-called modern races of man appeared so recently that they cannot be marked on a time line which includes the origin of life. The differences among Europeans, Africans, native Americans, Vietnamese, Eskimos, and indeed all peoples of the world are continually overdramatized. Since you have two parents, four grandparents, eight great-grandparents, and so on, and since there are about twenty-five years to a generation and thus four generations to a century, we are led to the conclusion that in forty thousand generations you would have accumulated 240,000 ancestors. This number is far greater than the total number of people ever to have lived. It exceeds by far the most radical estimates which anthropologists have made for the worldwide human populations one million years ago. If we assume that your ancestors were alive ten thousand years ago, this calculation can only mean that many, indeed most, of the relatives on your father's side were the same people as those on your mother's side. Moreover, it means that, whether Chinese or African, English or Dravidian your ancestors and your fellowman's ancestors were the same people.

Modern humans seem to have spread from west-central Asia to Borneo, Australia, and eastern Europe. By 35,000 years ago they were in western Europe; by 32,000 years ago they were in the Lena Basin in eastern Siberia and in Zaire in Africa; by 27,000 years ago they were in Namibia and perhaps northwestern America; by 19,000 years ago the Amerindians had reached Pennsylvania; and other Amerindians had spread to the furthest parts of South America by 10,500 years ago.

With a cranial volume of over 1,500 cubic centimeters and a propensity for tribalism, poetry, and cunning, all people today are far more like each other than any of us is like *Australopithecus afarensis* or *Homo erectus*. Yet the continuity from Pleistocene apemen to us can be traced in outline in the fossils. We reconstruct our scientific story from the jawbones, the cracked skulls of the victim prey, the stone tools, and the dazzling paintings. The evolution of man provides us with a beautiful example of classical Darwinism. In man, as in every species, changes in populations occur through time in response to selection pressures. Changes can be traced from primates that ate insects, fruits, and nuts to big-game hunters making pointed sticks and handaxes. But the gathering of more clues has led to greater rather than less difficulty in distinguishing distinct stages in the transition from small, nocturnal tree mammals to walking hominids. There is a certain continuity that goes from small Miocene apes running on the ground to Pleiocene australopithecines to *Homo erectus* to *Homo sapiens neanderthalensis* and to *Homo sapiens sapiens*.

During the Riss/Wurm interglacial period (the one prior to the interglacial period we are living in now), remains of fossil hominids become abundant. Especially in Europe, but in the Near East, southeast Asia, Zambia, and China as well, remnants of enigmatic women and men have been found. The "type locality" or place where these people were first recognized was Neanderthal, near Düsseldorf, a valley named after the seventeenth-century composer Joachim Neander.

The Neanderthalers can be told by their skulls; the average volume of known specimens actually exceeds those of modern people—but apparently in the wrong dimension. A rather chinless face, the Neanderthaler skull projects outward not at the top but in the rear. The Neanderthalers had short, stubby hands and probably strong, hairy grips. Neanderthalers, whatever else they were, were people. They were artists and poets and buriers of the dead.

Neanderthalers were replaced in Europe 35,000 years ago by people who physically were so like today's people that they would not be looked at twice on a New York City subway. Neanderthalers may have died out without leaving offspring. Some believe that Neanderthalers gradually became the modern Cro-Magnon people of Europe—the *Homo sapiens sapiens*. And some think that Cro-Magnon fought Neanderthalers and won, leaving hybrid offspring that became modern European man. Others have hypothesized that Neanderthalers, not very different from modern people, became us. Their heavy eyebrow ridges and jaws, their stooped posture and thick limbs, were the result of minor hormonal differences.

Just as small stature, barrel-shaped limbs, and thin arms characterized the Fuegians studied by Darwin and sketched by Robert Fitzroy during the 1830s voyage of the *Beagle*, so the Neanderthal physical type may have just been a "race" of the ancient human population. The hybrids that formed between Neanderthalers and Cro-Magnons may have led to the replacement of the extreme Neanderthal look by one more familiar. Lucy, if she dressed the part and you did not see her distinctly chimplike face, might be taken for a shrunken bag lady. *Homo habilis* would look severely arthritic and *Homo erectus* only slightly hunched. But Neanderthalers and Cro-Magnons, are, for all practical purposes, us. As we end the tale of the late Pleistocene peoples we recount the stories of ourselves. The question then becomes: If we have been around painting cave walls and designing artefacts for some 30,000 years at least, how long will it be until *Homo sapiens sapiens* becomes something else? When will a future descendent species look back on us as *their* naive predecessors?

Of course there is no definite answer. Invertebrate paleontologists who study the fossil remains of marine animals, and vertebrate paleontologists who study fish, amphibian, reptilian, avian, and mammalian fossils have concluded that distinguishable species of vertebrate animals which leave a clearly recognizable fossil record tend to persist about a million years. Invertebrate species seem to enjoy longer durations, about eleven million years. People, it may be argued, as the vertebrate species *Homo sapiens sapiens,* are at least a hundred thousand years old. Thus we have more than half a million years to go.

One may believe that mankind will live forever. But the concept that people are no longer subject to the evolutionary process is as irrational as a belief in Santa Claus or the Tooth Fairy. That our species will be replaced with none, one, or two descendant species within a million years or so is to be expected. It is expected on the basis of knowledge we have accumulated about *all* other species.

Our ancestors, with their special knack for culture, arrived on a microbial scene anciently begun. The planetary patina, the bacterial underlayer had long ago extended itself outward into weird macrocosmic proportions. The first peoples spread from their apparent point of origin in the Old World tropical Eden of Africa. Manapes expanded as *Homo erectus* and *Homo habilis* more than half a million years ago; their descendants, *Homo sapiens,* dispersed in huge numbers from central Asia more than 50,000 years ago. With their warm clothes and houses, and the vast ramifications of their abilities to think and work together in social collectives, our ancestors procreated and created, they dominated and predominated. They learned to select and cultivate plants, to outwit and teach each other. They became increasingly brainy and social mammals. They learned to sanctify with ceremony the transfer of information to their children. They did so with the abstract symbol systems of art, speech, magic, writing, religion, logic, and—starting at least 2,600 years ago with Thales in Greece—science.

Once they appeared, people sailed to the harsh New England coast; they trekked through sterile Arabian deserts; they rowed and climbed to the flats above the deep fjords of Scandinavia; they canoed to the inaccessible islands of the Pacific Ocean; and they eventually flew to the polar ice caps. Once human beings arose they spread everywhere. Yet there is little, in the end, beside our fecundity, persistence, imagination, and verbosity, that is very great or different about *Homo sapiens sapiens*. A sort of mammalian weed, with all our accomplishments and personality we are still the result of aeons of microbial recombination. With respiring mitochondria turning oxygen into energy and modified motility systems processing incoming sensory information, we resemble every other animal. We can boast that the cave paintings at Lascaux were rendered by that rare species who presumably came out of the trees and stood upright—on the face of the moon. But this is hero-worship, anthropocentrism. Certainly we can forgive it, but it would be more objective to put the same events another way: With respiring mitochondria and spirochetal secret agents dividing their cells, communities of the *microcosm* have alighted—if briefly—on the moon.

The extraterrestrial expansion of the ancient microworld has already begun. But this does not mean we are some sort of chosen species. Indeed, some scientists believe our fantastic recent success in populating the planet is a "sunset phenomenon": the bright lights before the inevitable end of the show. As the biologist A. Meredith suggests, the pattern of sudden appearance, expansion, and then disappearance in the fossil record has much historical precedent and is ominous. The lesson of the fossil past warns that superficially extremely successful life forms are often at the end of their biological tether. Historically, species just prior to their extinction often reproduce in considerable profusion. The many species of archeocyathids and trilobites in the Cambrian period, and of dinosaurs in the Cretaceous, are witness to this inauspicious process, which Meredith calls "devolution." As Charles Darwin realized, organisms adapt to their environment because of constant checks on their tendency toward unlimited growth. If they are not adapted they may decline in numbers and become extinct. But, according to Meredith, they may also become *too* adapted, multiply, deplete their resources, and *then* become extinct.

A microcosmic example of devolution would be microbes growing on a Petri plate. (Petri plates are round dishes with clear, transparent food permitting the investigator to see microbial colonies as spots even with the naked eye.) Fed on nutrient agar—bacterial food hardened with a gelatinous substance extracted from seaweed—microbes are often most prolific in the generations immediately preceding their collapse. Depleting all the nutrients in the agar and reaching the edges of the small laboratory dish, the multibillions of bacteria suddenly stop growing and die for lack of food and living space. For us, the world may be just such a Petri plate. Indeed, computer-enhanced satellite images of Spokane, Washington, show urban growth patterns similar to the growth of colonies of microbes. From the standpoint of Meredith's theory of devolution, it is easy to see that the implications of human population growth are not necessarily synonymous with progress.

Accepting that we are not higher than other organisms but equal to them—really just different recombinations of the same old microbial ancestors—we may want to come back and re-instill a bit of pride. Evolution is accelerating. The ancient bacterial microcosm seems to be on the verge of changes as major as any in its history. (Although presently mediated through humans, there is little guarantee we will continue to be its agents in the future.) To begin with, the technological jumps since World War II have been staggering. The Cambridge, Massachusetts, general systems theorist John Platt has studied and helped formulate the phenomenon of evolutionary acceleration.[48]

Platt believes life on Earth may be nearing "the greatest turning point in four billion years." He draws a series of stark comparisons, in a number of areas, between early life and present technology. Everywhere the impression is the same: the microcosm, as us, is gaining momentum so rapidly that there is no telling what may happen in even an extremely short span of geological time. First, let's take evolution itself. In evolution, the wanton effects of sexual crossing can now be mimicked and consciously directed by the techniques of molecular biology and recombinant DNA. Platt sees this as a significant advance over domestication and breeding and sees biotechnological ramifications "unfolding for thousands or millions of years."

In energy conversion, Platt traces a steady progression from photosynthesis 2,000 million years ago to "plant"-eating about 1,000 million years ago to fire about 400,000 years ago to farming and food production 10,000 years ago to the more modern methods of unleashing the energy locked up by photosynthesis in the form of coal and oil. Life has become most canny in the present era: photovoltaic cells offer a way of exploiting solar energy directly, while fission and fusion can reproduce the type of atomic reactions which take place in the sun to produce energy from matter in the first place.

Advances in life's ability to encapsulate itself and move into new habitats can also be identified. The path begins with the first cells living in water. Then multicellular life emerged onto land, protected by coverings such as shells, skin, and bark about 600 million years ago. The use of clothes and other artefacts enabled people to move to all climates, to establish cities on all continents within the last few thousand years, and to the move into the western "frontier" starting 600 years ago. The increasing sophistication of habitats has peaked in what Platt calls the "present transformation" of preliminary expeditions into space, the icy polar areas, and the Earth and oceans—all in appropriate capsules ranging from mine shafts to space vehicles to submarines. The absolute latest example of habitat complexity may be "Biosphere II"—a materially closed, energetically and informationally open ecosystem to be completed by 1992 near Oracle, Arizona. In addition to Earth studies, the structure is to serve as a prototype for spacecraft and stations that can recycle their own wastes and thus be freed of the need to receive materials from planet Earth.

Methods of travel, of course, have accelerated from cellular drift and spirochetes 3,000 million years ago to muscle systems such as fins, feet, and wings 300 million years ago. Distinct modes of human transportation probably started with boats 50,000 years ago, extending to horses and wheeled carts 5,000 years ago. Such methods were supplemented in the last couple of centuries by modern railroads, automobiles, and airplanes, and by jets and reusable rockets such as the space shuttle during the present transformation dating from the end of World War II. Tools and weapons have accelerated from the chemical phase of the early microcosm, to the teeth and claws of animal life about 600 million years ago, to the machines, guns, and explosives of the modern period, to the frightening remote-controlled and programmable nuclear weapons of the present transformation.

Once we get beyond the physical realm of artefact, or biological hardware, to what might be called biological information processing or software, the leaps become even more dizzying. For instance, an acceleration can be traced in perception systems.

Magnetotactic bacteria with tiny internal bars of magnetite

swam north (toward the magnetic north pole) in the northern hemisphere, and south toward the magnetic south pole two thousand million years ago. The bacterial chemical systems of detection and signaling of 3,500 million years ago were supplemented by cellular image formation and processing in protists some 1,000 million years ago. There is no doubt that at least two different dinomastigote protists, namely *Neodinium* and *Erythrodinium*, have eyes that see. With their analogous lenses and retinas, these single-celled eyes are on the lookout for their undersea enemies. Smelling, hearing, tasting, and the electricity-sensitive and echo-locating organ systems of certain animals were all well developed by 100 million years ago. Perceptual acceleration continued with speech in early humans about a million years ago, to writing at what is traditionally considered the dawn of civilization several thousand years ago, to telephone and radio in the modern phase, to a preliminary exploration of the electromagnetic spectrum including such things as radar, laser, and television.

Perhaps most profound are two final series of jumps life has made, in problem solving and storage, and in mechanisms of change. Since these two areas overlap, we will treat them together.

Problem solving begun 4,000 million years ago by way of mutating, recombining prokaryotic DNA chains. Natural selection, by preserving the bacteria and their descendants with the most effective responses to the environment, stored solutions to problems of overheating, drought, and ultraviolet radiation. The form of the storage was as informative sequences of nucleic acids and the capacity of these nucleic acids, RNA and DNA, to interact with protein in the immediate neighborhood. By transfering replicons, as explained in Chapter 5, the increasing quantity of information stored in the sequences of amino acids and base pairs of these long-chain biochemicals could be drawn upon, essentially from the time of life's beginning.

About 700 million years ago the evolution of the first nervous systems and brains set the stage for learning and thought, a more rapid means of problem solving that worked at the individual level. The new, neuronic problem solving worked not by Darwinian methods of individual death, or even by genetic exchange, but by methods associated with the psychologist B. F. Skinner—that is, by means of behavior modification. Instead of being stored in DNA, the variable behavior and selective reinforcement from the environment was stored in selective interactions among excitable cells or neurons, which respond directly to the environment. (As we have seen in Chapter 9, nervous systems may ultimately be based on bacterial locomotion.) From the early human phase until now the brain has tripled in size, growing at the approximate rate of 100 percent every million years.

A further jump in problem solving was science, which is usually attributed to the Hellenic Greeks of only 2,600 years ago. But the essence of formulating general laws about how the world works probably goes back as far as the first modern *Homo sapiens sapiens* 50,000 years ago. Science has become a social method of inquiring into natural phenomena, making intuitive and systematic explorations of laws which are formulated by observing nature, and then rigorously testing their accuracy in the form of predictions. The results are then stored as written or mathematical records which are copied and disseminated to others, both within and beyond any given generation. As a sort of synergetic, rigorously regulated group perception, the collective enterprise of science far transcends the activity within an individual brain.[49]

The practical application of science is research and development, which yields such things as the design of the atomic bomb, the launching of a man to the moon, the so-called strategic defense initiative of forming a laser shield over the United States, or space stations in Earth orbit with rivers and gardens in them. (The last two examples, of course, are still in the planning stage.) As Platt writes, "Today the new information-handling is at the heart of science, technology, warfare, banking, business, social accounting, and more and more of our daily lives. It is like a collective social nervous system for managing millions of our problems, and its importance for the long-run future may be as great as that of the first learning nervous systems." He characterizes large-scale research and development, the spending of up to one percent of the gross national product, as a process of "inventing evolutionary jumps, and it may be as big a jump as the original evolution of thought itself."

Looking at these examples of evolutionary acceleration, we cannot help but be awed. Certain phenomena which began in the dim abysm of the Archean Eon seem to have been expanding and gaining momentum for four billion years. These processes are overlapping, converging, feeding, and borrowing from each other. Like the symbionts of the microcosm, they may reproduce and recombine to become far more than a simple addition of their constituent parts. But only in retrospect will it be possible to determine whether the evolutionary acceleration observed by Platt was a prerequisite to greater things or merely a beautiful sunset phenomenon, Meredith's devolutionary dead end. We tend to take a middle road, believing that humanity is indeed just a phase, but an important one that will be included in future forms of living organization.

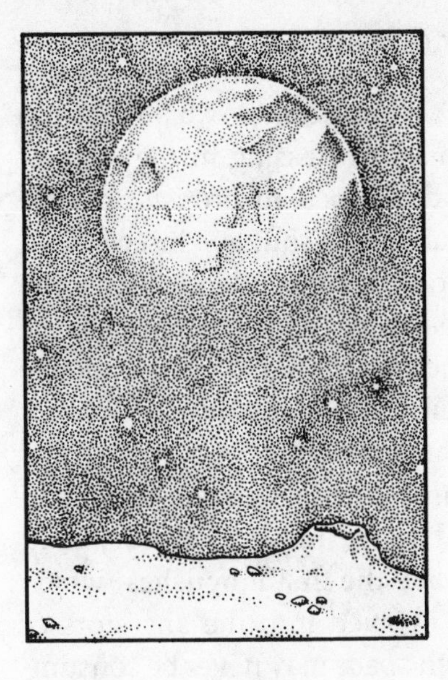

CHAPTER 13

The Future Supercosm

LIKE an adult whose personality traits were formed in a forgotten infancy, humanity and its place in history can be understood only as we explore and make sense of our cellular past. Plants and animals are patterns of nucleated cells. Though our nucleated cells come from the symbiotic bacteria of the microcosm, they form the beings which make up the macrocosm. Earth life will have to evolve to live on other planets, or even, perhaps, around other suns. And if we survive, we will certainly change, becoming part of the future "supercosm"—the hypothetical continued expansion of life from Earth into the solar system and beyond. The huge increase in area and resources will unleash life's potential, making the supercosm as different from our world as we are from bacteria.

Because we have a monopoly within the animal kingdom on high technology, human beings seem the most likely candidate to expand life throughout the solar system, if not beyond. But it is not a foregone conclusion that human beings will be the ultimate agents of the microcosm's expansion into space. For example, visual image processing in the form of eyes evolved many times; it developed in protists, marine worms, mollusks (such as snails and squids), insects, and vertebrates. Wings likewise evolved independently in insects, reptiles, birds, and bats: similar aerodynamic designs arose to meet the similar contingencies of the air. This tendency of organisms to evolve in similar directions despite the fact that they have different recent ancestors is called convergence. Convergence suggests that many kinds of beings will expand into space, just as many kinds have moved onto dry land and into the atmosphere. But like the first lungfishes, which came out of water but never evolved into the ancestors of land animals, our flirtations with space may never be consummated by continued life there. The presence of nervous systems and community behavior in many sorts of animals suggests that if human beings fail, other life forms will evolve to cart the primordial microcosm into space. If human beings become extinct—or if, like the horseshoe crab or lungfish, we just happily remain in our present habitats—the biota may, for a time, remain terrestrially confined. But remember it took humans only a few million years to evolve. Even if all anthropoids—all humans, monkeys, and apes—became extinct, the microcosm would still abound in those assets (e.g., nervous systems, manipulative appendages) that were leveraged into intelligence and technology in the first place. Given time to evolve in the absence of people, the descendants of raccoons—clever, nocturnal mammals with good manual coordination—could start their own space program. Sooner or later the biosphere is likely to expand beyond the cradle called Earth.

It is an illuminating peculiarity of the microcosm that explosive geological events in the past have *never* led to the *total* destruction of the biosphere. Indeed, like an artist whose misery catalyzes beautiful works of art, extensive catastrophe seems to have immediately preceded major evolutionary innovations.

Life on Earth answers threats, injuries, and losses with innovations, growth, and reproduction. The disastrous loss of needed hydrogen from the gravitational field of the earth led to one of the greatest evolutionary innovations of all time: the use of water (H_2O) in photosynthesis. But it also led to a tremendous pollution crisis, the accumulation of oxygen gas, which was originally toxic to the vast majority of organisms living on the planet. Nonetheless, the oxygen crisis 1,000 million years ago prompted the evolution of respiring bacteria which used oxygen to derive biochemical energy more efficiently than ever before. These bacteria were symbiotic and merged with other bacteria to form eukaryotic cells—which, becoming multicellular, evolved into fungi, plants, and animals. The most severe mass extinctions the world has ever known, at the Permo-Triassic boundary 245 million years ago, were rapidly followed by the rise of mammals, with their sharp eyes and large receptive brains. The Cretaceous catastrophe, including the disappearance of the dinosaurs 66 million years ago, cleared the way for the development of the first primates, whose intricate eye-hand coordination led to technology. World War II ushered in radar, nuclear weapons, and the electronic age. And the holocaust of Hiroshima and Nagasaki over forty years ago decimated Japanese industry and culture, unwittingly clearing the way for a new beginning in the form of the rising red sun of the Japanese information empire.

With each crisis the biosphere seems to take one step backward and two steps forward—the two steps forward being an evolutionary solution that surmounts the boundaries of the original problem. Not only meeting but going beyond challenges confirms that the biosphere is extremely resilient, that it recovers from tragedies with renewed vigor. Nuclear conflagration in the northern hemisphere would kill hundreds of millions of human beings. But it would not be the end of life on Earth, and, as heartless as it sounds, a human Armageddon might prepare the biosphere for less self-centered forms of life. As different from us as we are from dinosaurs, such future beings may have evolved through matter, life, and consciousness to a new superordinate stage of organization, and in doing so consider human beings as impressive as we do iguanas.

Needless to say, such a vision offers only metaphysical consolation. Barring direct impact by an atomic weapon, which of course is fatal, only 10 micrograms (that is, 10 millionths of a *gram*) of radioactive fallout—the debris which explodes into the stratosphere, blows in the wind, and later settles down—is needed to kill a person. Current estimates put Soviet and U.S. nuclear bomb arsenals at 10,000 *megatons* apiece. As the late inventor Buckminster Fuller showed by dropping tiddlywinks on a giant map spread across the ballroom floor of the New York City Sheraton Hotel, 5,000 bombs released at random on the globe would paralyze all the major population centers. And given present (as opposed to projected future) arsenals, a full-scale nuclear war is expected to deplete from 30 to 60 percent of the stratospheric ozone layer. The dust and smoke of city fires would rise up and surround the earth, first burning it but later leading to a devastating drop in worldwide mean temperature. Radiation could also cause worldwide plagues of AIDS-like diseases because of its compromising effects on the human immune system. Yet, with all that, we doubt that the overall health and underlying stability of the microcosm would be affected. The increase in radiation-induced mutations would have no direct effect on microbe evolution, since there has always been a huge reserve of radiation-resistant mutants to supply the evolutionary process. For example, *Micrococcus radiodurans* has been found living in the water used to cool nuclear reactors. Nor would the destruction of the ozone layer, permitting entry of torrents of ultraviolet radiation, ruin the microbial underlayer. Indeed, it would probably augment it, since radiation stimulates the bacterial transfer of genes.

In Theodore Sturgeon's science fiction story "Microcosmic God,"[50] a brilliant scientist, unable to "skim the surface of the future and just dip down" when he sees something interesting, invents life in the laboratory instead. The life form—"Neoterics," which repeat the evolutionary path from microbes to social and ultimately intelligent beings—has an unusually rapid generation time. The Neoterics evolve faster and faster as their creator submits them to more and more drastic tests. Finally, as they are organizing representation for themselves and communicating with the inventor through a system of his own devising, he gives them a final test, which they pass. When traces of aluminum oxide and other compounds are introduced into their chamber, the miniature beings metamorphose the molecules into solid pillars of pure metal. This prevents the walls of their habitat cage from closing in on them. Eventually the Neoterics isolate themselves from the outside world with an impenetrable and unknown material. Even their inventor can no longer contact them.

This story provides a metaphor for the present human predicament. Evolutionary trials and tribulations have stimulated the formation of creatures wondrous and strange, each, as we have seen, more surprising than the next. The natural challenge of glaciation that faced our ancestors, those tropical apes, has recently honed intelligence in its human form. Yet this intelligence, it seems, judging by the apocalyptic increase over time of our weapons of mass destruction, may ultimately prove self-limiting. The accelerating production of weapons—that in time may destroy the weapon makers themselves—is a kind of story very familiar in the history of the biosphere.

The accelerated nature of evolution in general and cultural evolution in particular makes it impossible to predict future evolutionary innovations, especially long-range ones. If we simply extrapolate current trends, we arrive not at the future but at a caricature of the present. For example, when the telephone was invented it was predicted that in the not-too-distant future every city and hamlet might have use of one. On the other hand, when helicopters were invented, there were those who saw the day when every suburban house would have one in its heliport. Respectable scientists, writing in formal journals with full citations to the professional literature and mathematical equations, predicted that the surface of the moon was covered with commercially exploitable levels of oil; they further stated that growing lichens that turned green in summer covered nearly an entire hemisphere of the planet Mars. Other scientists predicted dust layers so thick that landing on the moon would not be possible. So we will not pretend to have private knowledge of the future, but rather will discuss possibilities based on an awareness of the long-term past.

Beyond short-term technological fads are the long-term trends of life—extinction, expansion, symbiosis—which seem universal. We, the species *Homo sapiens,* will reach extinction, with or without a nuclear war. We may, like ichthyosaurs, seed ferns, and australopithecines, leave the annals of earth history without an heir, or we may, like choanomastigotes and *Homo erectus,* the respective ancestors of sponges and us, evolve into distinct new species.

No matter what our progeny evolves or devolves into, however, if it remains on earth it will eventually be scorched alive. By any astronomical reckoning, the sun has a lifetime of only about ten billion years. After it burns up all its primary hydrogen as fuel, nuclear reactions that break apart heavier atoms such as helium will take over. As it irradiates and expands into a red giant, our dying star will shine as it has never shone before. The luminous sun will generate immense heat, boiling and evaporating the oceans, destroying the atmosphere and melting the surface granitic and basaltic rocks. Our sun then is expected to run out of fuel. In its final stages the sun will gravitationally shrink hundreds of times, becoming a dense white dwarf, and finally a black dwarf, a tiny spent ember in the vast and fiery universe.

Could this natural nuclear disaster be the end of our form of earth life, the final unanswerable test of that genius molecule, DNA, and of human beings, should we still be alive, in whatever form, at that time? Is all our extraordinary microcosmic evolution to be swept away in the ordinary explosion of an ordinary star? Or is there an alternative, some ultimate scientific grounds for hope?

As the Earth is scorched, its oceans boiled to steam by the final outbursts of a waning sun, only that life that has wandered beyond the home planet (or protected itself in some way) will be salvaged. Today, Antarctic and Saharan organisms are the extremities of Earth life. They have appeared because the fat, watery, organic, and sunlight-bathed middle of life on Earth is so alive and well. American skyscrapers and subways, for example, are not independently generated; they are the outcome of a working, thriving primate culture, of a civilization fed by produce from the farmlands of the midwest and organized by computer connections from Los Angeles to New York. Only after much time and creative travail did *any* species emerge that could survive and reproduce in the face of extraordinary differences between their warm, wet insides and cold, dry outsides. The reader is one such example. Your insides are much more like the tropical shorelines of the Archean Earth than they are like the north temperate zones to which our species has cleverly adapted through the contrivances of clothing and shelter. Indeed with oil heating, philodendrons, spider plants, and steamy shower-baths, the effective environment of humanity is what it has always been: the African bush.

The prototypical African environment has found its way everywhere, from corporate high-rise Chicago meeting rooms, replete with their botanical frills, dracenas, and Boston ferns, to the insides of Eskimo igloos of the Canadian northwest, where burning seal blubber, warm fur, and communal bodies unite to the same effect. The *Homo sapiens* homeland—like that of the first cells—has made its merry way to the middle of the Atlantic Ocean in luxury cruisers. Bacterial offspring of the earliest cells entered human habitats and thus have even visited the moon. Dividing inside the spaceships, suits, and bodies of American astronauts, they give an inkling of how intricately life may have to remake its environment to survive in space. The exportation of the Edenic homeland of the human species has been cleverly brought about by our ability to extend by cumulative species experience—by culture—what was formerly transmissible only through genetic change. Yet the ironic end result of all our extragenetic intelligence is precisely the further preservation of the genes. At the present time intelligence seems to be a key mode of survival. If earth life is to survive the death of the sun it must extend microbial communities into a safer abode. Could our laudable autopoietic prowess, responsible for placing DNA into earth orbit, be extended still further into space?

Perhaps. There are many precedents of environmental transcendence and, after all, we are still playing the same old bacterial game. The bacteria never adapted to the severely cold, dry, airless, barren moon any more than we have adapted to the brutal, cold, damp dark of Scandinavia in winter. Rather, we and the bacteria of which we are evolutionarily composed have managed to manipulate these disparate environments in accordance with our needs. We help pulsate and expand the borders of Earth life. No matter what expenditure of energy is necessary either to carry with us, change, or create anew the ancestral environment, we do it. We bring the habitats of our predecessors into the homes of our future. The insertion of past dwelling places into new ones is an intense sort of conservatism, a deep-rooted refusal to change—or an agreement to change with the proviso of doing so in order to stay the same. Such a monomania for preservation may be just what is needed to rescue future organisms, and, with them, life itself, from the fate of an exploding sun. The future may then hold in store for us the final microcosmic metaphor—that of a planet whose ultimate fate is to "divide."

We can already see hints that the boundaries of life are growing. Populations, industries, universities, and suburbs have rapidly grown, but none has grown indefinitely without causing severe resource depletion and environmental transformation. Natural selection, which is just different rates of survival, whether in spirochetes or spider monkeys, can be emotionally frightening. Populations are beyond good and evil. They grow in response to the availability of space, food, and water. When too numerous, organisms either perish or transcend themselves.

If they transcend themselves they find new ways to procure room, carbon, energy, and water, which produces new wastes. The increasingly abundant production of new wastes tests those that made it. Life itself becomes a central figure creating its own problems and solutions. An example of such a problem would be pollution created by the use of compounds in the outer solar system as part of a program of resource acquisition by future corporations. Such toxic wastes might reach earth. On Earth, the solution, as in the case of the oxygen problem of the early microcosm, might be new organisms able to tolerate or make use of such wastes. This, in turn, would establish a living partnership that stretched millions of miles, from earth to the moons of Saturn.

Even to abstractly grasp the potential of life in the future, we must carefully look at life in the past. The dramatic evolution of man cannot be separated from the coevolution of our microbial ancestors, the bacteria constructing our cells and the cells of our food species of plants and animals. In coevolution over thousands of years partners change genetically. Inherited partnerships evolve together as new proteins and developmental patterns emerge. Ultimately the partners become totally dependent on each other, and it is no longer valid to consider them individuals. The agricultural grain corn provides a striking example of such coevolution, one that has occurred in human lifetimes, during the past few thousand years. No longer does corn wither naturally like the grasses from which it evolved: corn now must have its thick husk removed by human hands in each and every generation. Now its reproduction is tied to ours. It cannot complete its life cycle without us because it is part of us. Once teosinte, an inconspicuous, self-sufficient grass on the Mexican plateau, the plant has been selected by hungry peoples and grown for larger and larger kernels. It has become a major staple for humanity. The luxury of yesterday has become the necessity of today.

The fantastic increase in the human population depended on plants, and probably will continue to depend on them and their bacteria-derived chloroplasts if we are to move into space. It took a thousand hectares during the last interglacial period to support a single old-stone-age hunter. It takes over 10,000 times less space to support a modern Japanese rice farmer. Thus for every hunter that once roamed the island of Honshu there can now exist over 10,000 inhabitants in a Tokyo suburb (Table 3). Like the cells of the microcosm before us, human beings must coevolve with plants, animals, and microbes. Eventually, we will probably aggregate into cohesive, technology-supported communities that are far more tightly organized than simple or extended families, or even nation-states or the governments and subjects of superpowers. Inconspicuous seeds of the coming supercosm in space—paralleling the fleshy-finned fish that gave rise to all land

TABLE 3
ACCELERATION IN FOOD PRODUCTION

The amount of land needed to support human life has diminished dramatically from paleolithic (old stone age) blade makers 35,000 years ago to modern Japanese rice farmers—another example of evolutionary acceleration.

Human Culture	Land Needed to Support One Person	Time
Paleolithic hunters	1000 hectares*	35 thousand years ago
Neolithic cow-plow peoples	10 hectares	8 thousand years ago
Medieval peasants	0.67 hectares	1,000 years ago
Indian rice growers	0.20 hectares	100 years ago
Japanese rice growers	0.064 hectares	now

* 1000 hectares equals 10 square kilometers; 1 hectare = 2.47 acres

vertebrates, or the quirky system of meiotic sexuality that we have inherited from certain protists—may already be present on Earth. Such seedling systems could include various forms of political, economic, and technological organization.

Groups of organisms form new beings at higher levels of organization. Societies and populations are groups of organisms formed of members of the same species; communities are groups of organisms formed of members of different species (Table 4). Symbionts under certain pressures behave as single wholes. Individual bacteria became the organelles of nucleated cells; nucleated cells teamed up into many-celled "individuals" trillions of times their size. Larger beings—whose components are also beings—have been called "superorganisms."

Since symbiosis is the rule in evolution and organisms are always organized into communities of different species, no one species could make the transition to space alone. Humans seem well suited to help disperse the Earth-based biota, and they may occupy a prominent place in the supercosm—just as mitochondria, using oxygen inside the cells of plants and animals, helped such organisms settle the dry land. But for humans to play such a prominent role in the expansion of life into space, they must learn from the successful species of the microcosm. Humans must move more rapidly from antagonism to cooperation, and generally treat all species as fairly as a small farmer does his egg-laying chickens and milk cows. Unlike poaching rare animals for their pelts, or garishly displaying horned heads over a mantelpiece, or shooting birds for sport, or bulldozing rain forests, such fair treatment means living with other organisms. It means gradually forming superorganisms. Contrary to his hunting ancestors, the small farmer of today does not destroy a chicken or cow for a single feast, but cares for the animals, consuming their milk and eggs.

This sort of change from killing nearby organisms for food to helping them live while eating their dispensable parts is a mark of species maturity. It is why agriculture, in which grains and vegetables are eaten but their seeds always stored,

TABLE 4
HIERARCHY CHART: SIZES

Objects	Unit Size	Examples
atoms	angstroms	H,C,N,P
molecules	angstroms	H_2, NH_3, amino acids
macromolecules	angstroms	RNA, DNA, protein
organelles	nanometers	nuclei, ribosomes
cells	microns	bacteria, blood cells
tissues	microns	cartilage, epithelium
organs	millimeters	ovaries, leaves
organisms	centimeters	humans, daisies
populations	meters	cattle herd, locust swarm
communities	meters, kms	salt marsh, pond
ecosystems	kms	forest, coastal plain
biota	thousands of kms	sum of organisms
biosphere	thousands of kms	surface of Earth

NOTES:

Populations are groups of individuals, members of the same species, living in the same places at the same time. *Societies* are populations with members differentiated for specific tasks.

Communities are populations of organisms of different species living in the same place at the same time.

Ecosystems are groups of self-sustaining communities (members feed themselves, remove waste; no import or export of matter is needed for the autopoiesis of members). The smallest ecosystem known for certain is the *Biosphere.*

The *biota* is the sum of the living matter on Earth.

The *biosphere* is the place at the surface of the Earth where the biota exists; it extends from the top of the high mountains to the abyss.

is a more effective strategy than the simple gathering of plants. The trip from greedy gluttony, from instant satisfaction to long-term mutualism, has been made many times in the microcosm. Indeed, it does not even take foresight or intelligence to make it: the brutal destroyers always end up destroying themselves—automatically leaving those who get along better with others to inherit the living world.

The ancestors to the mitochondria of our cells were probably vicious bacteria that invaded and killed their prey. But we are living examples that such destructive tactics do not work in the long run: mitochondria peacefully inhabit our cells, providing us with energy in return for a place to live. While destructive species may come and go, cooperation itself increases through time. People may expand, plundering and pillaging the Amazon, ignoring most of the biosphere, but the history of cells says we cannot keep it up for long. To survive even a small fraction of the time of the symbiotic bacterial settlers of the oceans and Earth, people will have to change. Whether we move into space or not, we will have to dampen our aggressive instincts, limit our rapacious growth, and become far more conciliatory if we are to survive, in the long term, with the rest of the biosphere.

Even with an understanding of where we come from, our view of where we are going becomes blurred the further we look. But, as the visionary poet William Blake wrote, "What is now proved true, was once only *imagined*." There are many imaginable ways by which people could evolve into a species distinct from *Homo sapiens*. The simplest would not be by mutation at all but through sexual recombination of preexisting genes. Although all human beings are the same species, population extremes may be noted. A Pygmy woman, for instance, may not be able to give a Watusi man a baby because her pelvis is too small. This example illustrates the natural variety present in any species which may, over time, give rise to divergent species unable to interbreed because of outward changes resulting from altered symbionts, behavior, mitochondria, chromosomes, or nucleotide sequences in the DNA.

But cells can now be fused in forced fertilization and the simple accumulation of vast numbers of changes in DNA base pairs can now be engineered. The genetic "writings" of future biotechnologists ultimately may be new organisms. The use of sets of bacterial genes—or at least the funding for such use—has already become commonplace. Through biotechnology those pieces of DNA called plasmids are inserted into bacteria and thus quickly replicated. Genes coding for proteins, even human proteins, may be replicated via their association with plasmids. Pheromones, aphrodisiacs which modulate sexuality, or pituitary hormones, chemicals which control growth processes, may be manufactured by bacteria and later inserted into us, or into our plants and animals. With technological know-how, entire sets of genes, proteins, hormones, and other biochemicals are dovetailed, creating new species of microbes. Patents have been issued to those claiming to have developed new strains of life in the laboratory. An understanding of embryology and immune systems will help make it possible to clone cells at will into larger, more complex organisms. New species of organisms, as well as mythical monsters of folklore, may conceivably be manufactured by genetic engineers for entertainment purposes or as slaves. The Faustian industries of the future probably will overwhelm us with their ability to custom-make profitable life-forms.

The question of direct intervention in the processes of human evolution is a fascinating one. Presently it is being approached from several separate fronts: traditional natural selection (deforestation, animal and plant breeding) as well as biotechnology, computers, and robotics. Given that evolution accelerates, it must be only a matter of time before these approaches converge. Geologically speaking, we refer to exceedingly brief time periods, possibly even within our own lifetimes.

Computer science has been one of the most rapidly growing fields in the history of technology. From vacuum tubes to transistors and semiconductors, the information-handling elements of computers have miniaturized tens of thousands of times in only several decades. Their switching speed, the time needed to go from on to off in a binary code, has gone from twenty to a billion times per second. Psychiatric programs administered from behind partitions have fooled people into thinking they were communicating with human beings.

Intelligent machines already aid in the molecular design of new drugs. Computers efficiently transfer funds and keep huge files in memories far greater than those of any human beings. In offices and homes they have begun to replace writings on paper, or "hard copy," with the "soft copy" of magnetic disks and tape. We look forward to public information utilities that hook up to home disk libraries, TV screens, telephones, and printers. Worldwide access to information may decentralize government and demystify professional knowledge. The information revolution may even lead to a new age of participatory democracy. But it also may fragment society, dividing it into cloistered electronic homes and encouraging new forms of political exploitation and crime.

As computerized records, books, and other devices become commonplace because the raw, siliceous, and miniaturized components of computers are so inexpensive, society will transform. The trend for money to become increasingly electronic will continue. Education will become easier as teaching gadgets enter the market. Beyond the "paperless office," there will occur what the computer expert Christopher Evans has called "the death of the printed word."[51] Traditional printed books will become as extravagant—and as expensive—to future people as first editions or hand-printed manuscripts seem to us. Textbooks and trade paperbacks will appear to be immensely laborious undertakings. Each bulky mass of ink-spotted paper, such as the one you now hold in your hands, will take on the antiquated demeanor of, say, the Mainz Bible of Johann Gutenberg. Since the complex nature of future societies is bound to be dependent on and monitored by computer intelligence, social movements, financial transactions, and exploratory discoveries will be recorded in machine memories. Since retrieval of computer-stored events will be far more faithful than movie "re-creations" or historical novels, it will be possible to relive history and explore the past. Through technology, life's ancient ability to preserve the past in the present, its mnemonic fidelity, will vastly improve. This memory phenomenon, aided by cinema, written history, electromagnetic records, and other computer technology, is still in the process of acceleration.

Since silicon chips with thousands of bits of memory can pass through the eye of a needle today, microprocessors—tiny computers—are now lightweight enough to insert into machines, making them robots. Robots have great potential for the future. In 1976 the robotic part of the Viking spacecraft performed a task no human being could have done. Landing on the ultraviolet light-bombarded, frozen, and suffocating surface of the red planet, it stretched its mechanical arm, drew in a sample, and analyzed the dry and oxidized Martian regolith. Other robots are more mundane and economical. Metal robots with many arms fasten tires to cars with a productivity rate far in excess of their human counterparts. The assembly line itself is becoming assembled. In Japan robots make parts for other robots. More and more lifelike robots are becoming feasible as automated factories play an increasingly important role in the global economy.

As computers and machines come together in the new field of robotics, so robotics and bacteria may ultimately be united in the so-called "biochip," based not on silicon but on complex organic compounds, that is, an organic computer. Like plants performing photosynthesis, these manufactured molecules would exchange energy with their surroundings. But rather than turning it into cell material, they would turn it into information. The possibilities inherent in such a development are awesome. Such "living" computers could trade millions of hydrogen atoms per second and perhaps be integrated into conscious organisms. At this distance in the future the imagination is overwhelmed. The outcome of information exchange between computer, robotic, and biological technologies is not foreseeable. Perhaps only the most outlandish predictions have any chance of coming true.

What are some of the outlandish fates of *Homo sapiens* in the next couple of centuries? Let's explore a few of the many possible futures contingent upon biological activities. As we have seen, the nucleated cells of all animals, fungi, and plants contain genes packaged as chromosomes. Species are known to evolve by several means, including chromosomal rearrangements, the accumulations of mutations in DNA, and through symbiosis. Chromosomes undergoing heritable changes cause jumps in evolution larger than those caused by nucleotide base-pair mutations. Symbiotic leaps, like those of Jeon's amoebae, can, in a few generations, establish new species. Nothing prevents such modes of variation from operating on populations of human beings. Eventually, in fact, some of our descendants are bound to undergo chromosomal mutations or to acquire new symbionts.

To begin, let us envision chromosomal mutants, people with more than the two standard sets of chromosomes. Such polyploid people would be analogous to cotton, wheat, and showy flowers like carnations. The commercial strains of these plants are all polyploid. Polyploid people would probably be larger than their diploid relatives. They might be better adapted to life in low gravity.

In most cases polyploid mammals, those with extra sets of chromosomes, do not survive. Yet we do know that abrupt chromosomal changes, such as those involved in karyotypic fissioning, have led to many new species of mammals. Karyotypic fissioning is the name of a process in which chromosomes break apart at the kinetochores. Many species of Cenozoic mammals, compared with their ancestors, show half chromosomes, broken at the kinetochores. Neil Todd, publisher of the *Carnivore Genetics Newsletter,* believes that karyotypic fissioning is implicated in the evolution of dogs from wolves, pigs from boars, and even the manlike apes from their monkey ancestors. Combined with incest, karyotypic fissioning, in principle could lead to new—perhaps even evolutionarily accelerating—species of humans. The conquerors of the supercosm, if they are our descendants, or at least the descendants of some of us, are likely to have even more fissioned chromosomes than we.

Future humans may even be green, a product of symbiosis. An example of such a symbiotically produced species of human is "*Homo photosyntheticus,*" the imaginary cure to the heroin problem suggested by the algae expert, Ryan Drum. *Homo photosyntheticus* are green heroin or cocaine addicts who have had their heads shaved and their scalps injected with a thin layer of algae. Strung out under the lights, such green hominids do not have to be addicts, but Drum suggests that since they would be fed by their internal resources, if they were addicts they would no longer be a burden to society.

Evolution has already witnessed nutritional alliances between hungry organisms and sunlit, self-sufficient bacteria or algae. *Mastigias,* a Pacific Ocean medusoid, a peaceful coelenterate of the man-of-war type, helps its photosynthetic partners by swimming toward the areas of most intense light. They, in return, keep it well fed. This could happen to our *Homo photosyntheticus,* a sort of ultimate vegetarian who no longer eats but lives on internally produced food from his scalp algae. Our *Homo photosyntheticus* descendants might, with time, tend to lose their mouths. An analogous evolutionary fate has really befallen the red-tide protists *Mesodinium rubrum.* This ciliate bears a vestigial mouth, no longer used for feeding. It is well-enough fed by its internal populations of symbiotic algae. Furthermore, just as the skin of *Homo photosyntheticus* would tend to become very bald and pale, in order to let in enough light to its algae, *Mesodinium rubrum* is far more translucent than its fellow mesodinia of different species. *Mesodinium rubrum* also swims slowly backwards, basking in the light of the sun. In the same way, *Homo photosyntheticus* might be expected to be translucent, slothish, and sedentary.

Translucent flatworms of the species *Convoluta roscoffensis* contain green algae in between the cells of their tissues. On the beaches of Brittany and the English Channel, they are usually taken for green seaweed. Dark green *Convoluta roscoffensis* are "plant-animals." Adults have nonfunctional closed mouths. The algae not only live under the transparent skin of the worm, feeding it, but they recycle the worm's waste, uric acid. They take part of the uric acid molecule for themselves (the part with carbon and oxygen) and process the rest of it into further food for the worm.

In an analogous fashion symbiotic algae of *Homo photosyntheticus* might eventually find their way to the human germ cells. They would first invade testes and from there enter sperm cells as they are made. (This is hardly outrageous: insect bacterial symbionts are known to do exactly this. Some enter sperm, and some are transmitted to the next generation via eggs.) Accompanying the sperm during mating, and maybe even entering women's eggs, the algae—like a benevolent venereal disease—could ensure their survival in the warm, moist tissues of humans.

In the final stages of this eerie scenario we envision groups of *Homo photosyntheticus* lounging in dense masses upon the orbiting beaches of the future, idly fingering green seaweeds and broken mollusk shells.

We have suggested three possible paths of the evolution of humans. They are fanciful, perhaps, but the lessons of the past tell us that similar changes are inevitable. Our details may be absurd but change is certain. We can think of other peculiar possibilities. One is cybersymbiosis, the evolution of parts of human beings in future life forms. In this scenario, people are as crucial to the development of the supercosm as mitochondria or spirochetes were to the macrocosm. If we do transcend the fate of mammalian extinction and survive in an altered form we may persevere not as "individuals" but as remnants. We can imagine ourselves as analogous to spirochetal remnants: future forms of prosthetically pared people—perhaps only their delicately dissected nervous systems attached to electronically driven plastic arms—lending decision-making power to the maintenance functions of reproducing spacecraft.

• • •

Unfortunately for those who believe that humanity is the apotheosis, the culmination of life on Earth, the idea of machines that reproduce themselves is not a matter of scientific fantasy but a matter of fact in the present organization of the biosphere. Production, reproduction, and self-maintenance or autopoiesis are relative terms. If we consider reproduction to be life's most salient trait, and the biosphere life's most fundamental unit, then even the Earth cannot be considered alive, since it has not yet reproduced. Indeed, only DNA and RNA can directly replicate. Everything else—bacteria, girls, whales, weeping willows, McDonalds, and NASA space shuttles—reproduces indirectly through these molecules. Much molecular replication, cell growth, development, and construction is involved before two bacteria, two girls, two whales, two willow trees, two McDonalds, and two space shuttles appear in the biosphere.

Samuel Butler wrote that Darwin's idea of "rudimentary organs" could just as well be applied to human artifacts such as pipes and clothing. Surviving archaic quirks in clothing would include extra shirt buttons, pockets sewn closed, and decorative suspender loops. Butler even speculated that the little protruberance at the bottom of the bowl of his tobacco pipe might have descended from the old nonportable type, in which the protruberance, like the rim at the bottom of a teacup, served to "keep the heat of the pipe from marking the tabletop on which it rested." To Butler such rudimentary organs showed that, like mechanical artifacts, organisms were, if not designed, at least more creatively fashioned than could be explained by Darwin's theories. He also focused on the idea of machine development in order to satirize the blind enthusiasm with which his countrymen had greeted the industrial revolution.

All this is reflected in Butler's 1863 letter to the Christchurch,

New Zealand *Press,* which he entitled "Darwin Among the Machines." In this droll commentary, written four years after the publication of Charles Darwin's epochal treatise, *On the Origin of Species,* Butler compares the adaptive prowess of "mechanical life" to that of mere flesh-and-blood mortals. "There are few things of which the present generation is more justly proud," he begins, "than of the wonderful improvements which are daily taking place in all sorts of mechanical appliances." But what would happen, he wondered, if technology continued to evolve so much more rapidly than the "animal and vegetable kingdoms?" Would it displace us "in the supremacy of the earth?" Just as "the vegetable kingdom was slowly developed from the mineral," Butler reasoned, "and as in like manner the animal supervened upon the vegetable, so now in these last few ages an entirely new kingdom has sprung up, of which we as yet have only seen what will one day be considered the antediluvian prototypes of the race."

Butler conceded that machines were still governed by their makers, but, surveying the marvels of nineteenth-century technology, he wondered if that would always be true: "We are daily giving them greater power and supplying by all sorts of ingenious contrivances that self-regulating, self-acting power which will be to them what intellect has been to the human race." Already they were more efficient at converting raw materials into energy for work and generally required less maintenance than draft animals. Could it be long before machines were equipped with reproductive organs? "There is nothing which our infatuated race would desire more," he quipped, "than to see a fertile union between two steam engines."

Today, of course, the prowess of machines and their interdependence with human beings is far greater than it was in

Butler's time. Norbert Wiener, the founder of cybernetics, wrote in 1961 that "The idea of nonhuman devices of great power and great ability to carry through a policy, and of their dangers, is nothing new. All that is new is that we now possess effective devices of this kind. In the past, similar possibilities were postulated for the techniques of magic, which forms the themes for so many legends and folk tales."[52] From a biospheric view, such devices are just one of the microcosm's latest strategies for extending its domain beyond its present scale into the coming supercosm. The classification of machines as nonliving does not negate the fact that they reproduce, and reproduce with changes, as avidly as any viruses.

Agricultural contrivances with "viruslike" properties, such as tractors or harvesters, provide food that leads to further growth of the human population. Among this growing population are those in the agribusiness who design, develop, construct, and sell more tractors and harvesters and other devices to increase food output. By increasing the numbers of corn plants and people, tractors ensure their own reproduction. They are autocatalytic. Indeed, the potential of machines for exponential growth—the sort of evolutionary acceleration we noted earlier—far exceeds that of the bodies of humans. For instance, The World Future Society of Bethesda, Maryland, reported that the yearly percentage growth for the U.S. robot population in 1984 was 30 percent. During the same time period, the yearly percentage growth of the U.S. population was only less than 2 percent.

We are fond of labeling recent, large, adaptive, expanding—in a word, humanlike—populations of mammals "evolutionarily advanced." Even scientists tend to call organisms that combine large size, aggressive reproduction rates, rapid change, and recent evolutionary appearance "higher." Here we are claiming that by these stacked criteria machines are even more "evolutionarily advanced" than we are. They change their form at a far more rapid rate than any animals; witness the car, the phone, the copying machine, or the personal computer. Machines are able to survive more extreme environments than humans and other animals with central nervous systems. Machine generation time can be far shorter than that of humans. Machines outperform humans in information tasks such as arithmetic and printing. Machines have a greater range of mechanical energy at their disposal, including nuclear fusion, combustion, and photoelectric power.

That machines apparently depend on us for their construction and maintenance does not seem to be a serious argument against their viability. We depend on our organelles, such as mitochondria and chromosomes, for our life, yet no one ever argues that human beings are not really living. Are we simply containers for our living organelles? In the future humans may program machines to program and reproduce themselves more independently from human beings. Norbert Wiener believed that the reduction of oscillating systems of a given frequency by other oscillating systems of a different frequency to a common frequency constituted an exciting frontier of incipient electric reproductive processes. The most hopeful note for human survival may be the fact that we are presently as necessary for the reproduction of our machines as mitochondria are for the reproduction of ourselves. But since economic forces will pressure machines to improve at everything, including the manufacture of machines with a minimum of human work, no one can say how long this hopeful note will last. The computer designer John von Neumann unequivocally stated that sufficiently complex machines could be constructed to reproduce themselves in the absence of human help.

Another serious argument against the concept that machines are alive is that machines lack DNA and RNA, and are not made of carbon-nitrogen compounds in water. But beehives, the calcium phosphate of bones, and the exoskeleton of insects lack DNA and RNA as well. Living organization disappears during the process of analytical dissection. The debate over what is and what is not alive takes on new fascination as a close study of waste transformations and an atmosphere steeped in chemicals produced by life reveals that no clear line can be drawn between organisms and their environment, between what is "natural" and what is not. If life is defined as autopoietic reproducing entities based on reduced carbon compounds, then at first glance completely self-reproducing von Neumann machines can never be alive, since they are not based on carbon. Yet what is meant by "based"? Surely any invention of human beings is ultimately based on a variety of processes, including that of DNA replication, no matter the separation in space or time of that replication from that invention. This is not sophistry, a blurring of careful distinctions, or scientific reductionism, but rather what might be called "postanalytic reality."

Simply by extrapolating biospheric patterns, we may predict that humans will survive, if at all recognizably, as support systems connected to those forms of living organization with the greatest potential for perception and expansion, namely machines. The descendants of *Prochloron,* the chloroplasts, retained a much higher rate of growth inside plant cells than did *Prochloron,* their free-living green bacterial relatives patchily distributed in the Pacific Ocean. Analogously, human beings in association with machines already have a great selective advantage over those alienated from machines.

The future evolution of the supercosm may be compared to the evolution of nucleated cells from coevolved communities of bacteria in the microcosm more than a billion years ago. Life may continue to expand via DNA-people-machine-based ("technobic") entities. Given the phenomenon of evolutionary acceleration elucidated by John R. Platt, it may, in staggeringly short time periods, penetrate vast regions of the galaxy. Relative to former evolutionary advances—the conquest of dry land and winged flight into the atmosphere—galactic habitation may occur nearly instantaneously. Perhaps within the next few centuries, the universe will be full of intelligent life—silicon philosophers and planetary computers whose crude ancestors are evolving right now in our midst. From a long-term outlook, our position vis-à-vis the future is not conducive to human chauvinism. Trace fossils of machines already exist in the solar system beyond the Earth. From 1976 until the shortage of NASA funding caused their connection with earth to be severed in the early 1980s, the Viking orbiters and landers periodically surveyed the lonely stillness of the Martian landscape. Other machine extensions of the biosphere, in the form of spacecraft orbiting various celestial bodies, primarily Earth, are still very much "alive." They are connected to the system and in fact are less vulnerable to external threats than are we flesh-and-blood beings.

Extending living trends, decoded from readings in the rock record of the ancient history of life, into the very short-term geologic future, we can say that mammalian extinctions and replacements, including our own, will continue. So will the appearance of new life forms both mechanical and organic. Soon the former may outnumber the latter, and technology, catalyzed by future crises, may prove as important to the next round of evolutionary innovation as clones of nucleated cells were for the appearance of technology. Given the high pitch of acceleration that exists in the present transformation, we may see hints of such wholesale renovations in the basic structure of life within the next few decades.

In one of his comments to the editor of *The Press,* Butler wrote: "We treat our horses, dogs, cattle, and sheep, on the whole, with great kindness, we give them whatever experience teaches us to be best for them, and there can be no doubt that our use of meat has added to the happiness of the lower animals far more than it has detracted from it; in like manner it is reasonable to suppose that the machines will treat us kindly, for their existence is as dependent upon ours as ours is upon the lower animals." Man, he consoled, "will continue to exist, nay even to improve, and will be probably better off in his state of domestication under the beneficent role of the machines than he is in his present wild state." Samuel Butler was kidding, in order to show the preposterousness of a purely mechanical view of life. But he also may have been a visionary who saw an important aspect of the coming supercosm: the use of technological machinery in carrying the wet, warm environment of the pre-Phanerozoic microcosm into a future as fascinatingly nonhuman as the past.

We occasionally still hear scientists and the press express fears that we will infect planets reached by our space probes with microbes from Earth. But this is just what will have to be done if people are to live for extended periods of time in space. To live economically in space stations, or on the Moon, Mars, or beyond, people must import the biotechnology of the microcosm. Of course this is not simply infecting planets; it is "fertilizing" them, inoculating them with the correct kinds and numbers of microbes so that a supportive habitat may develop. Disease often reflects the excessive, opportunistic growth of microbes normally present; indeed, the death of such microbes may not cure but exacerbate the illness, since some "disease" microbes serve the salutary purpose of restraining still other microbes. For example, gram-negative bacilli, normal gut bacteria which nonetheless can cause pneumonia in susceptible newborns, may be prevented from growing by the introduction of certain strains of streptococcus bacteria. The health of ecosystems is comparable: it depends on populations of different organisms, on the metabolism, growth, and coevolution of the associated biota. The idea that space stations can be landscaped into self-supporting Gardens of Eden in the absence of microbial communities is simply ludicrous. Such stations would be sterile "tin cans" in space.

Preliminary studies at NASA's Ames Research Center in Moffett Field, California, show the importance of biological complexity in the creation of stable living systems. Experiments with microbes enclosed in plastic boxes and placed in the light suggest that greater numbers of species—in other words, more complexity—tends to enhance stability and photosynthetic food production. Although the supercosm may be created by our better-adapted descendants, huge numbers of interacting beings in tightly organized communities may require millions of years to produce stable ecosystems capable of supporting human life off the Earth. We probably cannot carelessly infect planets even if we tried; they must be inoculated. Seeding the supercosm would seem to require a rare combination of scientific playfulness, a pioneering spirit, and a microbial "green thumb."

Although its continuance may depend on its ability to cut loose from the earthly umbilical cord and extend beyond the solar womb, life itself so far has proved to be immortal. As a conspicuous planetary phenomenon, life's self-organization seems to violate the second law of thermodynamics, which implies that the universe is becoming less orderly, that it is running down. But if life evolved from the universe, how can it be based on a set of principles apart from those of the universe? What is life? Paradoxically, we have had to wait until the end before trying to answer this question.

Ancient peoples saw nature as *animus*, as moving spirits, animals, and gods. More recently life on Earth was thought to exemplify the creative processes of God. In Cartesian and Newtonian thinking life is seen "objectively" as a part of universal physical movements. Life turns and meshes like gears in some great mechanical clock; organisms act and react like balls on a billiard table. Here, all is stimulus and response, cause and effect. Today we hear new, computer-age analogies: amino acids are a form of "input," RNA is "data processing," and organisms are the "output," the "hard copy" controlled by that "master program," that "reproducing software," the genes. In this book we have held to a somewhat different and more abstract view of what life is. This new view of life is less facile than the obviously outmoded magical, religious, and scientific views. Life, a watery, carbon-based macromolecular system, is reproducing autopoiesis. The autopoietic view of life is circular. Life is a metabolic machine which not only reproduces but fiercely stores and uses information in order to resist breaking down.

Architect-philosopher Buckminster Fuller has called the biosphere "Spaceship Earth"—whirling around the sun as the sun itself whizzes and bobs through space. But the analogy of the Earth as a spaceship suggests that we are the planetary drivers, which we are not. Another concept is offered by physician Lewis Thomas, who has compared the biosphere to a cell, which suggests orchestration and unity, and to an embryo, which suggests future growth. John Platt has extended the simile of the earth as an embryo, pointing out that the biosphere may currently be passing through some sort of birth canal. Although he does not want the analogy taken uncritically, he notes that in late pregnancy hormones produce prebirth changes, cycles of pain begin, and the mother realizes that the converging events leading to birth cannot be stopped or even slowed down. Such imagery suggests that the Earth's slow gestation period may soon be over, leading to rapid growth in the supercosm. Finally—although our list is hardly exhaustive—the freelance atmospheric chemist James Lovelock sees life best represented by a self-supporting environmental system which he calls Gaia.[53]

Gaia—named by the novelist William Golding (at Lovelock's request) after the ancient Greek goddess of the Earth—works in mysterious ways. Gaia, the superorganismic system of all life on Earth, hypothetically maintains the composition of the air and the temperature of the planet's surface, regulating conditions for the continuance of life. While the intricate network of biological relations by which life does this is still not well known, the fact that the biota monitors portions of the planetary surface is as well established as the fact that our body keeps itself at a constant temperature. Gaia thus may keep atmospheric nitrogen and oxygen, so important to life, from degenerating into nitrates and nitrogen oxides, into salts and laughing gas which could halt the entire system. If there were no constant, worldwide production of new oxygen by photosynthetic organisms, if there were no release of gaseous nitrogen by nitrate- and ammonia-breathing bacteria, an inert or poisonous atmosphere would rapidly develop. Under the reactive influence of scores of lightning bolts sparking the Earth's atmosphere every minute, the Earth would be no more hospitable to life than is acid Venus. On Earth the environment has been made and monitored by life as much as life has been made and influenced by the environment.

The striking thing about our blue, white-flecked planet is

that the idiosyncrasy of life, with its incredible diversity and peculiar biochemical unity, continues. Because we are constrained to communicate in standard English it is difficult for us to grasp the idea of the definition of life as a reproducing autopoietic system. Yet, according to Lovelock's idea, which he calls the Gaia hypothesis, the biota itself, which includes *Homo sapiens,* is autopoietic. It recognizes, regulates, and creates conditions necessary for its own continuing survival.

The fossil record supports the notion that the Earth's surface has been continuously regulated since the earliest widespread appearance of microbial life. The Gaia hypothesis that the temperature and reactive gas composition of the Earth's atmosphere are actively regulated by the biota was developed by Lovelock while he was working for NASA on ways to detect life on Mars. He found that gases which when tested in simple chemical systems react quickly, easily, and completely to make stable compounds coexist in our atmosphere. These gases seem to remain aloof in apparent disregard for the laws of standard equilibrium chemistry. Lovelock found the chemistry of the Earth's atmosphere so persistently bizarre that it could only be attributed to the collective properties of organisms, namely to the biota. Indeed the biota, especially the microbiota, constantly produces prodigious amounts of these reactive gases. By looking for such unlikely gas mixtures in the atmospheres of other planets with spectroscopes mounted on telescopes, he could, he thought, detect alien biospheres without leaving the Earth. Turning his attention to Mars, Lovelock discovered it to be in a balance entirely understandable on the basis of physics and chemistry alone. He postulated the absence of life on Mars by observing the absence of the Gaian phenomenon. But in 1975, NASA, all ready to land on the red planet, was not eager to publicize Lovelock's simple solution to the age-old problem of life on Mars.

But nothing was lost. The Viking spacecraft launched in 1975, arrived as two landers and two orbiters in 1976. The biological experiments carried aboard and soft-landed on that planet's surface were spectacularly successful. They showed definitively that there is no evidence for life on the red planet. Lovelock's work provided a basis for understanding the results. Furthermore, his analysis led to a new look at the biosphere. As great as the mystery of life elsewhere in the universe was the mystery of life on Earth. Why does the Earth have an atmosphere so far from what is expected on the basis of chemistry? Given the fact that oxygen gas forms 20 percent of the atmosphere, the relative disequilibria of methane, ammonia, sulfur gases, methyl chloride, and methyl iodide, among others, are enormous. According to chemical calculations, the quantities of all of these gases, which react so readily with oxygen, should be so minute as to be undetectable. But they are present and they continue to be present whenever anyone seeks them. Indeed, there is more than 10^{35}—ten to the thirty-fifth power or one followed by 35 zeros—times as much methane in the Earth's atmosphere as there should be considering the quantity of oxygen available to react with methane. Other gases, such as nitrogen, carbon monoxide, and nitrous oxide, are only ten billion, ten, and ten trillion times more abundant, respectively, than they should be given chemistry alone.

A further enigma concerns the temperature of the Earth. The laws of physics seem to make it inescapable that the sun's total luminosity—that is, its output of energy as light—has increased during the past 4 billion years, maybe as much as 50 percent. Yet evidence from the fossil record indicates that the Earth's temperature has remained relatively stable, the mean value hovering at about 22 degrees centigrade—about room temperature—despite the freezing temperatures expected from a puny early sun. Not only did it appear that life has been regulating the composition of gases on a worldwide scale, it also seemed that the temperature of the planet itself was under some sort of continuous control. What was this great hidden thermostat?

Rejecting mystical solutions, Lovelock theorized that the biota—and especially the bacterial microcosm—from its earliest appearance on the planet must have been regulating its environment on a global scale. Life forms react to perturbing geological and cosmic crises; they resist assaults on their individual integrity as long as possible; and these individual actions lead to a general maintenance of conditions favorable to collective survival. (This does not mean that there were never fluctuations: there were. For example, judging from the broad extension of fossil tropical forests during the Cretaceous, the planet was significantly warmer during the time of the dinosaurs; both before and after there were extensive ice sheets covering much of the planet. But in between and after such periodic fluctuations, the planet stabilized, never broiling like Venus or freezing like Mars.)

If the biota had not responded to major external perturbations such as the rise in solar luminosity or meteoritic impacts as devastating as nuclear bombs, we would not be here now. Life, Lovelock concluded, is not surrounded by an essentially passive environment to which it has adapted. Instead, it makes and remakes its own environment. The atmosphere, like a beehive or a bird's nest, is part of the biosphere. Since carbon dioxide is transformed into cells and can also be used to control the temperature of the atmosphere, it seems likely that one way life regulates the temperature of the planet is by modulating the atmospheric level of carbon dioxide.

Some scientists protest Lovelock's analysis. The idea that all life forms a body that responds physiologically to environmental threats and insults in order to assure its survival does not jibe with accepted notions of Darwinian evolution, which depend on *competition* among struggling organisms. Assuming Love lock is right, how does the struggling mass of genes inside the cells of organisms at the earth's surface *know* they are confronted by crises? How were microbes able to overcome the rise of oxygen levels in the atmosphere? How could they act together as a concerted, coordinated entity to meet such crises? W. Ford Doolittle, the molecular biologist who has done seminal work on the molecular biology of plastids, protested this notion that nature could be, as he put it, "motherly." The Oxford University zoologist Richard Dawkins likened the Gaia hypothesis to the "BBC Theorem"—a pejorative reference to the television documentary notion of nature as wonderful balance and harmony. Dawkins could not conceive of the evolution of worldwide Gaian control mechanisms without the universe being "full of dead planets whose homeostatic regulation systems had failed, with, dotted around, a handful of successful, well-regulated planets of which Earth is one."[54]

To answer such criticisms Lovelock devised some mathematical models. The most spectacular one, called Daisyworld, involves a mythical planet that can be covered only with black and white daisies—and an occasional daisy-munching cow. The daisies represent two species, both of which grow in patches and cover up to seventy percent of the planet within specific temperature ranges. Both grow not at all in the very cold, slowly in the cold, more quickly in the warmth, and not at all—in fact they die—at oppressive temperatures above 45 degrees centigrade.

Lovelock, later working with Andrew Watson at the Marine Biological Association in Plymouth, England, found that white and black daisies could act as a giant thermostat, stabilizing the temperature of an entire planet simply by growing. The phenomenon is not mystical but synergistic, the unexpected result of a complex system.

You can imagine how Daisyworld works. Take a planet of white and black daisies circling a star which is becoming slowly but steadily more brilliant. At first, since the sun is cool, not many daisies will grow. As the sun becomes a bit warmer, however, patches of both colors of daisies sprout and flourish. But as the black daisies open their flowers and produce more offspring, because they are dark and sunlight-absorbing, they keep light from being reflected back into space. The growth of clumps of black daisies, absorbing heat, has the effect of warming up an otherwise cool planet. But soon the local surroundings become oppressively hot and begin limiting the growth of black daisies in the immediate vicinity. Raised local temperatures lead to an overall higher temperature than expected on a daisyless world. Clumps of white daisies begin to grow. This, however, leads to an increase in planetary reflectivity, or albedo, as the white surface of the daisies' petals reflects light into space, which results in a return to planetwide cooler temperatures. Black daisies crop up. But the sun, meanwhile, continues to become hotter, and white daisies, reflecting heat, continue to bloom and cool off the planet. In short, the sun gets hotter, clusters of white daisies spread, Daisyworld gets cooler, which renews the growth of overheated patches of black daisies, which again become overheated, which again leads to more favorable conditions for white daisies, which again cool off Daisyworld by increasing its albedo, and so on until the sun becomes a red giant and burns up all the daisies. Yet—within certain temperature limits—the daisies act very simply as a thermostat, keeping the world livable despite a potentially deadly increase in the amount of solar energy reaching the planet.

To a remarkable degree the mute flowers regulate the temperature of the planet within the narrow range required for their survival as the sun relentlessly heats up.

In models more like the real world it is the growth, metabolism, and gas-exchanging properties of microbes, rather than daisies, that form the complex physical and chemical feedback systems which modulate the biosphere in which we live. Living organisms, through their effect on water and clouds, have an immense modulating influence on the Earth. Tiny algae floating on the sea, as just one example, can hypothetically send the world into an ice age simply by growing more rapidly in northern latitudes. In making their calcium carbonate shells, dying, and sinking to the bottom of the sea, they remove carbon available for making carbon dioxide; since carbon dioxide is a "greenhouse" gas which acts like an invisible blanket to let in light and trap it as heat, less atmospheric carbon dioxide can translate into colder temperatures. But with colder temperatures, fewer algae grow, less carbon dioxide is removed from the air to make shells, and the planet becomes tropical. So the feedback loops are so intertwined that massive death of marine algae, coupled with water erosion of carbonate rocks, a process which releases CO_2 into the atmosphere, can also cause an atmospheric *rise* in temperature.

Indeed, in 1979 and 1980 European researchers analyzed fossil air trapped in polar ice and found that about 20,000 years ago, at the peak of the last ice age, carbon dioxide was only two-thirds as prevalent as it was at the beginning of the industrial revolution. Just before humans became agricultural and formed the first civilizations, carbon dioxide rose to its preindustrial value. The rise in carbon dioxide and temperature 12,000 years ago took place in less than 100 years, and cannot be fully explained by traditional geophysical or geochemical processes such as tectonic activity or weathering.

Such an abrupt fluctuation could only have come from life. Lovelock believes the sudden death of a substantial proportion of marine algae probably caused this quick rise in worldwide temperature, an environmental transformation which ultimately allowed people to emerge from their cave dwellings and populate the Earth.

Over time, the biota has built up elaborate control systems of which we are only now becoming vaguely aware. The multitude of sensory systems of living organisms, their capacity for metabolism and exponential growth, and the extraordinary diversity of interacting life on Earth are themselves enough to account, in principle, for environmental modulation on a global scale.

But environmental modulation of a similar fashion also operates on a smaller scale. Even on the far smaller scale of individual animals, temperature regulation involves more than a simple, single feedback system. Let us do a thought experiment in which a person—like the biota, a collection of cells—becomes confronted with a sharp drop in ambient temperature. His first sort of response would be the most recently evolved: the high-tech response. This would consist of turning up the thermostat, plugging in an electric heater, or even paying a utility bill through a modem on his home computer. Though these may be increasingly common forms of temperature regulation, they are so recent they are still the most fragile of the various feedback systems. Underlying them are the "low-tech" responses to cold temperature threats: wrapping up in blankets and putting on extra clothes. This sort of technology, inherited from the custom of trapping and hunting cold-climate animals and borrowing their thick furs and skins, is about 100,000 years old. Sewing, a major improvement in this original technology, was one that, judging from archeological reports of wooden sewing needles, may have helped eastern peoples traverse the Bering Strait to North America. The feedback cycle of clothes is simple: when the weather becomes cool, clothes are put on; when it warms up, they are taken off. Temperature regulation behavior appeared in people far earlier than fossil-fuel heating systems and still is more prevalent. All peoples on earth today wear some kind of clothes.

If we continue to stress our subject we will induce in him a still older and more dependable means of temperature regulation: the nontechnological, behavioral feedback systems. These responses date back to some 200 million years ago and consist of running around, rubbing limbs together, cuddling, and curling up into a foetal position when cold. When threatened with heat, mammals like us respond by an opposite behavior. We spread our limbs and search for shade. In general we become less active. All mammals share this sort of temperature control mechanism, which depends on having a sufficiently complex nervous system, the basis for learned behavior. As we get closer to the original microcosm, the feedback systems become even more predictable, basic, and reliable.

Still older than behavioral systems are the strictly physiological sorts of control. As the surroundings cool, the blood vessels of mammals autonomically move away from the surface of the skin as the muscles in the walls of the blood vessels contract. Farther from the skin, the blood supply to the vital organs increases and they are protected. Frostbite follows. Fingers, toes, and other extremities become icy and numb. If still stressed, the abandonment of extremities occurs. Nose and eartips, fingers, and toes are sloughed off. Sweating, the opposite sort of response, evaporates water to cool off the body. These physiological responses to temperature are even more ancient and ingrained than the others. They are perhaps as old as animals themselves (about 600 million years).

If, in our thought experiment, we continue to stress our

man with cold we push the autopoietic system to its limit and lay bare the ancient genetic means of temperature control. If the environment becomes cold beyond the limits of tolerance of the man, he dies and leaves no (further) offspring. If the stress of the cold continues, his entire population and community freeze to death without offspring. New populations and communities, however, replace the old ones, some of them with more effective methods to combat cold. Only different or mutant organisms tolerant of the severe climatic condition survive. An enormous selection pressure is put upon those organisms that can ameliorate the effects of the local cold surroundings.

This has always been the way of the world. If stress is severe enough, only tolerant organisms survive. In other words, when it is too hot, cells die. When it is too cold, cells die. When it is just right, cells leave many offspring. But "just right" differs with each kind of life. Darwinian natural selection is the ultimate ancient Gaian feedback system upon which all the newer technological and behavioral ones are based. Today, if you are cold you first turn up the heat, then put on a sweater, then start shivering to generate heat. If still threatened you go into a state of sleep and lowered metabolism, and, if the threat is relentless, you die. But your individual death is part of larger systems of environmental stabilization. Before you died you increased the ambient temperature; and by dying and leaving no similar offspring, you lessened the chances that future cold spells will destroy life by clearing the way for cold-adapted organisms to reproduce.

Planetwide living systems of temperature and atmospheric regulation can only be guessed at. From a planetary perspective, however, they do not seem to be in a dainty natural balance on the verge of collapse. They are robust. The most important ambient control systems are time-tested, gas-producing, albedo-changing, microcosmic institutions far tougher and older than the burning of heating oil and the use of home thermostats. As for the future, our species may be like those flourishing black daisies that grow so fast they optimize the environment for others even as they roast themselves to death. Each individual, population, or species is an option that is exercised only under favorable conditions. If catastrophe strikes, as it regularly has in life's history, some options will no longer be viable. But their expiration, in the form of death or extinction, makes the biosphere as a whole stronger, more complex, and more resilient. (This, of course, has nothing whatever to do with human progress or well-being. There is no evidence for progress in the fossil record, only for change and expansion.)

Furthermore, it seems that most of the prokaryotic options have not yet expired. Neither the existence of species nor species extinction is a property of bacteria. Although individual bacterial death is continuous, fierce pressures on the moneran kingdom as a worldwide gene-exchanging enterprise led to the rapid exchange of natural biotechnologies, enormous population growth rates, and in general the ability to persevere with metabolic talents intact even during the most severe planetary crises.

Only with a full scientific exploration of Gaian control mechanisms can we expect to implement self-supporting living habitats in space. If we are ever to design closed ecosystems that replenish their own vital supplies, we must study the natural technology of the Earth. Inhabiting other worlds, making it possible for us to stroll through gardens upon, say, Mars, is a gigantic project only thinkable from a Gaian perspective. We should know our roots in the microcosm before we go out on that limb, the supercosm. But whether people carry the primeval environment of the ancient microcosm into space or die trying, life does seem tempted in this direction. And life, so far, has resisted everything but temptation.

NOTES

1. Eukaryotes include the familiar plant and animal kingdoms, as well as the less familiar fungi and protoctist kingdoms. The informal term *protists* refers to the microbial, often single-celled, members of the kingdom Protoctista. Protoctista include amoebae, ciliates, malarial parasites (and, in general, the protozoa), diatoms, seaweeds (and, in general, the algae), slime nets, water molds, slime molds, plasmodial plant parasites, and other more obscure organisms that don't fit into the other kingdoms. Nearly 250,000 species of protoctista, grouped into about fifty phyla, are estimated to be alive today. The other three eukaryotic kingdoms, in order of their evolution, are Animalia: animals, which develop from embryos that form after the fusion of a sperm with an egg; Fungi: molds, mushrooms, yeasts, rusts, puffballs, and related organisms that develop from spores; and Plantae: mosses, liverworts, ferns, cone- and flower-bearing plants that develop from embryos surrounded by maternal tissue. The fifth, and earliest kingdom of living things to evolve, is the kingdom Procaryotae, composed entirely of bacteria (prokaryotes). (The several names for bacteria—monerans, prokaryotes, germs, etc.—come from the traditions of their separate study within different fields of science. Natural history, botany, microbiology, medicine, agriculture, and zoology have maintained extremely different traditions of identifying, naming, and classifying the microbes.) The term *microbe* has no specific meaning in taxonomy or evolution, and is equivalent to *microorganism,* meaning any organism primarily seen through a microscope. All prokaryotes and many eukaryotic organisms, such as protists

and fungi, are also microbes in that they are beyond the resolution of the human eye. Since *microorganism* and *microbe* are synonyms, we generally use the more biological and less medical term *microbe* in this book.

2. Some biologists still do not believe in the symbiotic origin of mitochondria, chloroplasts, and other eukaryotic organelles. They are, however, increasingly in the minority. It is hoped that the weight of the evidence presented in this book will convince biologists—as well as everybody else—of the need to view life as a symbiotic phenomenon.

3. Charles Darwin, *The Variation of Animals and Plants under Domestication*, Vol. 2 (New York: Organe Judd, 1868), p. 204.

4. Francis Crick, *Life Itself: Its Origin and Nature* (New York: Simon & Schuster, 1981).

5. Cited in W. Kaufman, 1980, *Discovering the Mind*, Vol. 3, *Freud versus Adler and Jung*. McGraw-Hill, New York, p. 467.

6. Forum, *Harper's Magazine*, Vol. 280, No. 1679 (April 1990), 37–48.

7. J. L. Hall, Z. Ramanis and D. J. L. Luck, Basal body/centriolar DNA: molecular genetic studies in *Chlamydomonas, Cell* 59(1989):121–32.

8. D. Sagan, *Biospheres: Metamorphosis of Planet Earth* (New York: McGraw-Hill, 1990); paperback edition New York: Bantam Books.

9. L. Margulis and E. Dobb, "Untimely Requiem," review of *The End of Nature* by W. McKibben, (New York: Random House, 1989), *The Sciences* January/February 1990, 44–49.

10. M. McFall-Ngai, "Luminous Bacterial Symbiosis in Fish Evolution," Chapter 25 in L. Margulis and R. Fester, eds., *Evolution and Speciation: Symbiosis as a Source of Evolutionary Innovation* (Cambridge, Mass.: MIT Press, 1991), pp. 380–409.

11. Atsatt, P., "Fungi and the Origin of Land Plants," Chapter 21, in L. Margulis, and R. Fester, eds., *Evolution and Speciation: Symbiosis as a Source of Evolutionary Innovation* (Cambridge, Mass.: MIT Press, 1991), p. 301–316.

12. First line of poem 1440 (circa 1877), in T. H. Johnson, ed., *The Complete Poems of Emily Dickinson* (Boston/Toronto: Little, Brown & Co. 1890, 1960), p. 599–600.

13. Last stanza of poem 1400.

14. Steven Weinberg, *The First Three Minutes* (New York: Basic Books, 1977).

15. The remaining one percent of the dry weight of organisms is made up of rarer but equally indispensable elements, including zinc, potassium, sodium, manganese, magnesium, calcium, iron, cobalt, copper, and selenium.

16. For the relation between autopoiesis and the origins of life, see G. R. Fleischaker, "Origins of Life: An Operational Definition" in *Origins*

of Life and Evolution of the Biosphere, Vol. 20 (Dordrecht, Netherlands: Kluwer Academic Publishers, 1990) pp. 127–37.

17. In 1979, Stanley M. Awramik of the University of California at Santa Barbara discovered spectacularly well preserved multicelled filamentous and fatter, rounded microstructures in the rocks of the Warrawoona Formation in Australia that are also some 3,400 million years old. If the fossils are the same age as the surrounding rock, this find proves that even more complex structures existed even earlier than we have learned so far. But radioactive dating cannot be done directly on sedimentary rocks—the only rocks that harbor fossils—because they are composed of particles of different ages. Dates are estimated from surrounding volcanic rock. By 1991 the Archean dates of the Australian microfossils were confirmed.

18. Carbon, in nature, comes as several chemically nearly identical but physically different forms. The differences are due to the number of neutrons in the atom's nucleus. Most carbon around us is in the stable form of C^{12}. Less than one percent is in the stable isotope C^{13}, and an even smaller amount is in the unstable radioactive form C^{14}. Since in the process of photosynthesis organisms favor the use of C^{12}, the ratio of C^{12} to C^{13} in natural materials has been used to give clues as to whether or not certain carbon deposits originated by photosynthesis.

19. The equation to calculate bacterial growth is 2^n, where n equals the number of generations. Thus 3 generations/hour × 48 hours = 2^{144} bacteria.

20. Spores are a way to survive long periods of desiccation or other adverse conditions. A case forms around the cell's genetic and protein-synthesizing material and alters chemically, after which the rest of the cell disintegrates. The spore germinates again when conditions become favorable. Bacterial spores are known to lie dormant and ready for life for decades. Though the human life span prohibits direct testing, it is quite possible that dry spores can last for hundreds, perhaps even thousands, of years.

21. See MacLyn McCarty, *The Transforming Principle: Discovering that Genes Are Made of DNA* (New York: W. W. Norton, 1985).

22. Sorin Sonea and Maurice Panisset, *The New Bacteriology* (Boston: Jones and Bartlett, 1983), p. 22.

23. The idea that life not only uses and is made of its environment but also regulates the chemical composition of the atmosphere (keeping it away from chemical equilibrium) is called the Gaia hypothesis.

24. Sonea and Panisset, p. 85.

25. Quoted in Frederick Turner, "Cultivating the American Garden," *Harpers Magazine*, August, 1985, 45–52. For further details, see *Earth's Earliest Biosphere: Its Origin and Evolution*, ed. J. William Schopf (Princeton, N.J.: Princeton Univ. Press, 1983). Schopf's book provides technical

analysis and differing viewpoints on life's early phase. For a less technical primer of the often-overlooked time from life's beginning to the origin of eukaryotic cells some 1,000 million years ago, see Lynn Margulis *Early Life* (Boston: Jones and Bartlett, 1982).

26. James Lovelock, *Gaia: A New Look at Life on Earth* (New York: Oxford Univ. Press, 1979), p. 69.

27. Chet Raymo, *Biography of a Planet* (Englewood Cliffs, N.J.: Prentice Hall, 1984), p. 72.

28. W. H. F. Doolittle and Carmen Sapienza, "Selfish genes: the phenotype paradigm and genome evolution," *Nature* 284 (April 17, 1980), 601–3.

29. 1 meter of DNA per chromosome times 46 chromosomes per cell equals 46 meters of DNA per cell; multiplied by 10^{12} cells per body equals 460 quadrillion meters of DNA in a person's body. The Moon is only 239,000 miles away.

30. Oral communication. For details, see R. Klein and A. Cronquist, "A consideration of the evolutionary and taxonomic significance of some biochemical micromorphological and physiological characters in the Thallophytes," *Quarterly Review of Biology* 42 (1967), 105–296.

31. For references and details of Jeon's work, see his chapter, "Amoeba and X-bacteria: Symbiont Acquisition and Possible Species Change" in *Symbiosis as a Source of Evolutionary Innovation, Speciation and Morphogenesis,* eds. Lynn Margulis and Rene Fester (Cambridge, Mass.: MIT Press, 1991).

32. For details of Axelrod's conventions of competing nonzero-sum game participants, see Robert M. Axelrod, *The Evolution of Cooperation* (New York: Basic Books, 1984).

33. *Thermoplasma* grows optimally in an atmosphere about 5 percent oxygen and dies long before oxygen in its surroundings reaches 20 percent, the current level.

34. Some eukaryotes living symbiotically in the anaerobic interior of other organisms have themselves lost their mitochondria. (Their structure and reproductive behavior clearly mark them as eukaryotes, not as bacteria that evolved without mitochondria.) Still others lost them and then acquired new symbiotic bacteria as mitochondria surrogates—symbiosis within a symbiosis. These surrogates aren't exactly new mitochondria, as they function in an environment much lower in oxygen than what is needed for mitochondria to survive. They seem, however, to serve the same functions.

35. 1 angstrom = 1 ten-billionth of a meter. 10,000 angstroms = 1 micron. 1 micron (micrometer) = 1 millionth of a meter. Bacterial cells are usually about one micron across. One thousand microns make a millimeter and ten thousand microns make a centimeter. A sphere of about 500 microns can be seen—as a tiny grain of sand—with the naked eye. The subvisible

microcosm, objects seen with various kinds of microscopes, thus extends from hundreds of microns to about 0.001 micron or 10 angstroms, the distance across the DNA helix.

36. David C. Smith, "From extracellular to intracellular: the establishment of a symbiosis," *Proceedings of the Royal Society* (London) 204 (1979), 115–30.

37. Since flagella, as mentioned earlier, also refers to the whiplike projection attached to the rotary motor of bacteria, it is better to call eukaryotic flagella undulipodia, and leave the name flagella itself solely for use with reference to bacteria. See Margulis, L. and Sagan, D., *Origins of Sex* (New Haven, CT: Yale University Press, 1990).

38. Charles Darwin, *The Origin of Species by Means of Natural Selection or the Preservation of Favored Races in the Struggle of Life*, first edition 1859, Pengiun Classics edition 1968, reprinted 1981 (Harmondsworth, Middlesex, England: Penguin, Limited), J. W. Burrows, editor, p. 453.

39. M. A. Sleigh, "Origin and evolution of flagellar movement," *Cell Motility* 5 (1985), 137–73. This issue contains the proceedings of a seminar held under a U.S.-Japan Cooperative Science program, called "Fundamental Problems of Movement of Cilia, Eukaryotic Flagella and Related Systems."

40. D. Wheatley, a British anatomist, has written an entire book, *The Centriole: A Central Enigma* (Amsterdam, New York, and Oxford: Elsevier Biomedical Press, 1983) in which he beautifully reviews a great deal of cell biological literature but leaves the enigma of the title unsolved.

41. Albert Einstein, "Letter to Jacques Hadamard," in Jacques Hadamard, *The Psychology of Invention in the Mathematical Field* (Princeton, N.J.: Princeton Univ. Press, 1945), reprinted 1954 (New York: Dover Publications, Inc.), pp. 142–43.

42. John von Neumann, *The Computer and the Brain* (New Haven: Yale Univ. Press, 1958), p. 82.

43. Quoted in Alan Moorehead, *Darwin and the Beagle* (New York and Evanston, Ill.: Harper & Row, 1969), pp. 259–61.

44. "The Damned Human Race," in Mark Twain, *Letters from the Earth*, ed. Bernard DeVoto (1938; rp. New York: Harper & Row Publishers, 1962), pp. 215–16.

45. Calder's book, *Timescale: An Atlas of the 4th Dimension* (New York: Viking Press, 1983) is required reading for anyone interested in the expanding microcosm. Only in recent years has a comprehensive view of how life evolved on our planet been developed. There is more than one history of the earth, and there is constant scientific discussion and revision of details. Calder compares this new field of life's history to early mapmaking, and its practitioners to the first charters of the lands and oceans. His book,

which "attempts to put in order the significant events between the origin of the universe and the present," is one of the first atlases of this new discipline of chronography, mapping time. It demonstrates that history is more than a collection of dates, clay pots, and the names of kings. Indeed, the traditional written history taught in textbooks is only the tip of the iceberg. True history is prehistory, our living heritage. Those wishing to pursue this new field further should see *Biography of a Planet* by Chet Raymo (Englewood Cliffs, N.J.: Prentice Hall, 1984). This book, written at an introductory level, is a fine guided tour of life's development since its inception over three billion years ago.

46. William Irwin Thompson, "On Food-Sharing, Communion, and Human Culture," a sermon delivered at the Cathedral Church of St. John the Divine, November 1, 1981.

47. Thompson, one of the most thoughtful social critics of our time, has written *The Time Falling Bodies Take to Light,* (New York: St. Martin's Press, 1981), a delightful book in which he examines our species' penchant for mythopoiesy or the creation of myths, which we then take very seriously.

48. John R. Platt, "The Acceleration of Evolution," *The Futurist,* February, 1981.

49. For a brilliant analysis of the extraordinarily social route needed for someone's observation to become a scientific fact, see Ludvik Fleck's *The Genesis and Development of a Scientific Fact,* originally written in German in 1936. A beautifully translated and annotated version was published as an inexpensive paperback by the University of Chicago Press in 1979.

50. This story can be found in *The Science Fiction Hall of Fame,* Vol. 1, ed. Robert Silverberg (Garden City, N.Y.: Doubleday, 1970), p. 87–111.

51. Christopher Evans, *The MicroMillennium* (New York: Washington Square Press, 1981), pp. 112–121. The book projects the short-, middle-, and long-term future of computers.

52. Norbert Wiener, *Cybernetics: Or Control and Communication in the Animal and the Machine* (Cambridge, Mass.: MIT Press, 1961), p. 176.

53. For a lively discussion of the Gaia hypothesis by its inventor, see Lovelock, *Gaia: A New Look at Life on Earth* (New York: Oxford Univ. Press, 1979), his *Ages of Gaia* (New York: W. W. Norton, 1988), and his chapter in *Global Ecology: Towards a Science of the Biosphere,* eds., R. Fester and L. Margulis (Boston: Academic Press, 1989).

54. Richard Dawkins, *The Extended Phenotype* (San Francisco: W. H. Freeman, 1982), p. 236.

INDEX

ABOUT THE AUTHORS

Lynn Margulis is a Distinguished University Professor at the University of Massachusetts at Amherst and author of *Symbiosis and Cell Evolution* and *Early Life.* She is also the coauthor of *Five Kingdoms,* with K. V. Schwartz, and *Mystery Dance* and *Origins of Sex,* with Dorion Sagan. Professor Margulis is a member of the National Academy of Sciences.

Dorion Sagan is author of *Biospheres* and coauthor of *Mystery Dance, Origins of Sex* and *The Garden of Microbial Delights.*